# MEMORY
## MAKES THE
# BRAIN

The Biological Machinery that Uses
Experiences to Shape Individual Brains

# MEMORY
# MAKES THE
# BRAIN

## The Biological Machinery that Uses
## Experiences to Shape Individual Brains

### Christian Hansel

The University of Chicago, USA

**World Scientific**

NEW JERSEY · LONDON · SINGAPORE · BEIJING · SHANGHAI · HONG KONG · TAIPEI · CHENNAI · TOKYO

*Published by*

World Scientific Publishing Co. Pte. Ltd.
5 Toh Tuck Link, Singapore 596224
*USA office:* 27 Warren Street, Suite 401-402, Hackensack, NJ 07601
*UK office:* 57 Shelton Street, Covent Garden, London WC2H 9HE

**British Library Cataloguing-in-Publication Data**
A catalogue record for this book is available from the British Library.

MEMORY MAKES THE BRAIN
The Biological Machinery that Uses Experiences to Shape Individual Brains

ISBN 978-981-122-880-3 (hardcover)
ISBN 978-981-122-881-0 (ebook for institutions)
ISBN 978-981-122-882-7 (ebook for individuals)

For any available supplementary material, please visit
https://www.worldscientific.com/worldscibooks/10.1142/12058#t=suppl

Typeset by Stallion Press
Email: enquiries@stallionpress.com

*In loving memory of my father*
*Joachim Hansel*

# Contents

# Acknowledgments

My first thanks go to my wife, Anne Scarlett. Anne teaches business writing and oral communication at Columbia College and DePaul University in Chicago and, with that expertise, has helped to improve my writing and bring structure to it. Anyone communicating in a non-native language knows all too well how difficult it is to ensure that the nuance comes across as intended. I feel blessed and grateful that with patience and competence, Anne has worked on these initial corrections. I would not have felt as comfortable with anyone else doing this for me.

The book reflects on a wide range of neurobiological implications of memory that, at times, go beyond my personal expertise in the field of cellular physiology. My appreciation and gratitude go to my colleagues that have read specific chapters of this book and offered constructive feedback. These colleagues are Drs. Michael Brecht (Humboldt University, Berlin), Winrich Freiwald (Rockefeller University, New York), Elizabeth Grove (University of Chicago), Kimberly Huber (UT Southwestern, Dallas), Bobby Kasthuri (University of Chicago), Siegrid Löwel (University of Göttingen) and Clifton Ragsdale (University of Chicago). I also want to thank John Goldsmith, Professor of Linguistics and Computer Science at the University of Chicago, for discussions about the role of symbolic representation and associative learning in language acquisition. This is a tremendously interesting field with relevance to the brain sciences. Without John's guidance and feedback, I would not have been able to write about it.

This book is a monograph, but several chapters are based on reviews or perspective articles that I have written in collaboration. I would like to acknowledge these colleagues for their much deserved credit: Drs. Nicolas Brunel, Henrik Jörntell, Masanobu Kano, Peggy

Mason, Claire Piochon, Dana Simmons and Heather Titley. It has been a pleasure and wonderful experience to work with you all.

I would like to warmly thank my editors at World Scientific Publishing, Rachel Field and Sandhya Devi, for their great work and confidence in me. Without our frequent communications and their enthusiasm about the project, this book could not have been completed. I would also like to thank the production team around Subash M Swamy for their excellent craftsmanship.

I have had two academic mentors, Drs. Wolf Singer (Max-Planck-Institute for Brain Research, Frankfurt) and David Linden (Johns Hopkins University), who deeply inspired me to become not only a neuroscientist with expertise in a limited subject, but to always strive to assume a wider, possibly philosophical perspective on all things concerning the brain. Both are brilliant scholars and have found ways to express their unique thoughts: By refusing to restrict themselves to publishing work in traditional academic journals, they found their own voices.

My parents, Barbara and Joachim Hansel, and my siblings, Babette, Stefanie and Jochen, deserve thanks that are hard to express in words. Over all these years, they have supported my decision to first, find my education as a postdoctoral scholar, and then, my academic home overseas, in Baltimore and Chicago, far away from home, and thus removed from the type of frequent interaction that typically defines a family. I remind myself every day that this is made possible by their unwavering love and understanding.

Finally, as at the beginning of these acknowledgments, I would like to thank my wife Anne. This time for her never ending support, love and encouragement, without which I would not have been able to gather the courage and motivation to write this book.

# Introduction

Each of us has a brain that is entirely unique. Individuality arises from a genetic history that we share and from an organismal history that is comprised of our personal experiences. At about the time when I was making plans for this book on memory mechanisms in the developing and adult brain, my father died, leaving behind a family that was not ready to let him go. There have been so many moments while writing this book, when I was pausing and thinking about what an integral part of my history he is, the history that shaped me. This book is inspired and guided by the memory of my father.

Brain plasticity enables learning. When writing about learning, I will use the term 'learning' in a way that slightly differs from how it is used by educators or psychologists, namely, the acquisition of knowledge or skills. That definition originated in early 20th-century behaviorism, which aims at an understanding of learning that is based on measurable outcomes in the form of observable behaviors. It adheres to a traditional view of the brain as a black box that receives input and generates output. Learning then changes what happens in the black box to alter output upon receiving the same input. I will call this 'behavioral learning.' From a brain scientist's perspective, a wider, biological definition is more useful. Here, the changes that happen within the black box are of interest. The neural networks within the brain, the brain circuits, adapt, both functionally and structurally, to experiences. For the purpose of this book, I will call this phenomenon 'circuit learning.' In the absence of circuit learning, behavioral learning cannot happen. That said, circuit learning can also occur without a measurable outcome in organismal behavior. In those instances, changes at the microstructural level occur that ultimately benefit the organism, without expressing

1

themselves behaviorally. This distinction is important. In the context of brain maturation, I will also talk about autism as an example of a brain developmental disorder, where the formative processes that this book is about are altered. Autism is not a learning disorder in the sense of behavioral learning. However, pathological alterations of circuit learning can be identified. I will return to the topic of autism in Chapters V and VI.

Brain development is guided by nature and nurture principles. The history that is reflected in our brain circuits is a genetic history (nature) as much as it is a history of environmental influences (nurture) that impact maturation and learning throughout all stages of our lives. Remarkably, this dualistic nature of organismal 'memory' with both its phylogenic (phylogenesis describes the development/evolution of a group of organisms) and ontogenic components (ontogenesis is the development of the individual organism from inception to maturity) was already pointed out in 1904 by the German zoologist Richard Semon.[1] Semon introduced the term 'engram' to emphasize a shared commonality between the genome and behavioral learning: experience is 'engraphed' in biological substance, leaving a permanent trace that reflects an organism's history. Semon erred when claiming that individually acquired engrams could be inherited, a position that unfortunately dominates his work. But his more general argument that the genome is shaped by the cross-generational experience of selection pressure[2] introduces a useful engram concept that helps us to understand the contributions of phylogenesis and ontogenesis to brain development. In Chapter VII, I will return to the engram concept and to Richard Semon, one of the most remarkable but also tragic figures in the history of the biological sciences.

When we look closely at the development and maturation of our mental capacities, the brain assumes the role of a theater stage. We

---

[1] Semon, R. (1904). *Die Mneme als erhaltendes Prinzip im Wechsel des organischen Geschehens* (Leipzig: Verlag von Wilhelm Engelmann).

[2] Semon's work clearly shows Lamarckian influence, but he also admired Charles Darwin and his theory of evolution by natural selection. 'Cross-generational experience' here widely refers to the transfer of genomic information within an entire species and the many generations that it encompasses.

are witnessing a drama that has all elements of a classic tragedy; death and the struggle for survival go hand in hand, and by the end, the characters no longer resemble their former selves—they have evolved. So, too, it is with the central characters in brain maturation— the sites of physical contact between neurons called synapses. The word 'synapse' was suggested in 1897 by the English physiologist Charles Sherrington. At any moment in time, large numbers of neurons communicate with one another via synapses, and these synapses undergo plastic changes to improve and to learn. When we peer inside the brain to inspect the machinery within, we discover a highly dynamic mesh of neurons and synapses. An adult human brain has about 100 billion neurons, and well over a trillion synapses. These synapses constantly 'learn,' the micro-circuits in which they are embedded 'learn,' the brain 'learns'—enabling the entire organism to adapt. Why not distinguish *circuit* or *brain plasticity* from behavioral learning, restricting the word *learning* to its usage in the vernacular? *Plasticity* describes the ability for lasting change. *Learning* is a consequence of this change. Neural circuits that undergo plastic change maintain related information that they use to alter their local output signaling. According to this view, learning takes place at a substrate (here, a neural circuit) level, as much as it takes place at a behavioral level. Learning is an organismal response and can best be described as a hierarchical process that involves synapses, circuits as well as the entire organism—the behaving animal.

I mention brain development for a reason. As I will show later in this book, synaptic plasticity in adulthood is similar to synaptic plasticity during postnatal brain development, except that the magnitude of changes in synaptic weight[3] is less dramatic. In fact, it would be fair to describe brain development as a learning process and adult learning as the continuation of brain maturation. In short, regardless of differences in the degree of plasticity, the same forms of synaptic plasticity—the strengthening and weakening of synaptic weight known as potentiation and depression, respectively—occur

---

[3] Also used: *synaptic strength* or *synaptic efficacy*.

throughout the entire lifespan of the brain after birth. There is, however, an aspect of plasticity that is different from developing and adult brains that can be summarized in the following statement:

*The Preeminent Form of Synaptic Learning in the Developing Brain is Depression, and in the Adult Brain it is Potentiation*

This statement does not contradict what I outlined before. Bidirectional synaptic plasticity—that is, the ability to evoke an upward or downward change in synaptic weight—exists throughout development and persists into adult life. However, there are differences in criticality at the different stages of postnatal life. The developing brain is tasked with eliminating the excess synaptic connections that are formed during early postnatal development. Many of these synapses are not well integrated into their local neural network. Others operate efficiently and contribute to the function of their local network. The optimization of circuit architecture is critical during this developmental period and relies on the activity-dependent *pruning* of synaptic connections that are no longer required. The term *pruning* is quite appropriate for this process. It resembles the pruning that a gardener has to perform on trees and shrubs to prevent overgrowth and the waste of limited energy resources that are available for an individual plant. Synaptic pruning is the critical step during early postnatal brain development. For this reason, it is appropriate to describe the underlying depression mechanisms as the preeminent form of synaptic learning in the developing brain—even though some synapses stabilize and potentiate, and even though synapse formation precedes the pruning period.

The range of adaptive processes during childhood and adolescence is broad. It includes such basic adaptations as optimizing sensory perception. It also includes complex adaptations such as the shaping of brain circuits under the influence of sociocultural experiences, which, as argued by Wolf Singer at the Max-Planck-Institute for Brain Research in Frankfurt, are essential for generating the perception of oneself.[4] The dramatic events during this maturation phase and the

---

[4]Singer, W. (2019). A naturalistic approach to the hard problem of consciousness. *Front. Syst. Neurosci.* 13, 58.

rules governing pruning will be described in Chapters I and II. In the adult brain, synapses have reached a mature state and undergo experience-dependent changes in synaptic input weight to store information content in specific learning contexts. New information is encoded through a strengthening of synaptic weight termed 'long-term potentiation' (LTP) at synapses that are activated by it. Information storage by usage is an important concept in learning.

When I was a teenager, my family took long summer vacations in the Austrian mountains. With my mother as the driving force behind all outdoors activities, and my father following somewhat reluctantly, we took many hikes together as a family. Mother upfront, father trailing, and somewhere in between and around them were their four children, often accompanied by friends. Among the many memories I have of these family hikes, one stands out in particular. It is a memory of an unusually large and beautiful butterfly—I believe it was a swallowtail—perched on a stone, warming up in the midday sun. There is much to be said about this seemingly unremarkable memory in the context of perception, brain maturation, and learning. At birth, in ways that are poorly understood, our brains already comprise a functional but immature neural network—a default state of connectivity that enables us to experience basic perceptions. A butterfly is perceived as what it is and is distinguished from other objects in the visual world. However, you need a father (or other figure who cares that you get this right) to help you make sense of it, to explain that a butterfly is 'harmless,' 'good,' and 'beautiful,' and that for various reasons it is 'bad' to hurt it. This is an example of how we learn to categorize and associate descriptive labels and evaluative narratives with objects.

My father was a physician, and many of my normative experiences were influenced by the humanistic ideals and educational values of the *Bildungsbürgertum* during the 20th century. I fondly remember how my father made me run into the little town where I grew up to buy a walking stick for one of his patients. This was a small act of kindness, nothing overly remarkable, but it left an impression on me and certainly contributed to my appreciation of social norms. My favorite memories are of the times when we played music together, particularly around Christmas. My mother plays the cello, my sister the violin, and with several family members playing the piano (my other sister, my brother, and me), and my father singing with plenty of enthusiasm, we always had a small family orchestra available, good enough to master the Christmas classics. We still do this, now without my father. But in my childhood, this time together certainly imprinted me in so many wonderful ways, providing normative musical and social experiences, without which I would be a different person. It is such experiences that shape our brains. When they occur during early childhood, the biological machineries involved become part of a larger drama of synaptic strengthening and pruning events that ultimately create our individual personalities. It is difficult to describe, even schematically, what happens within our brain circuits when formative events stretch out over long periods of time. But what about acute memories, such as that of the butterfly spotted during a family hike? Why can I recall this memory so many years later?

A general scheme of the events involved in memory storage and recall would look like the one shown in Figure 0.1. First, there is the

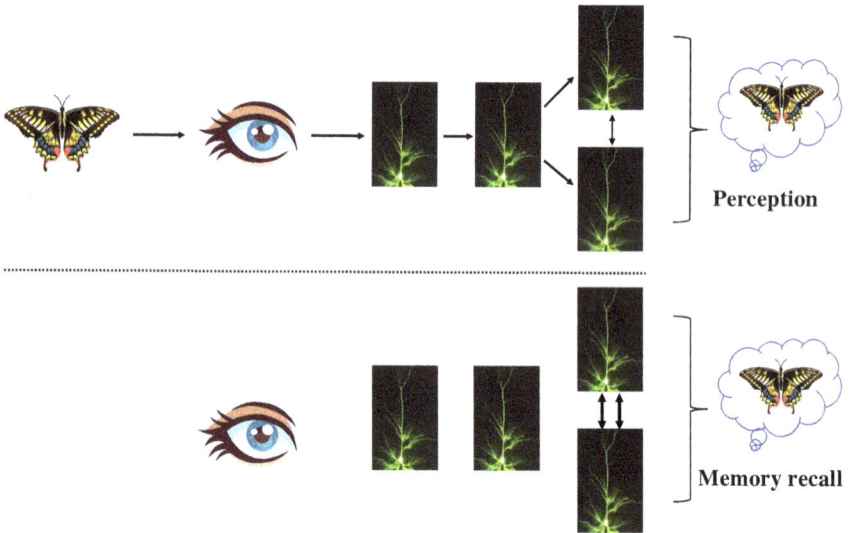

**Figure 0.1:** A memory engram emerges from the strengthening of synaptic connections between neurons ('engram cells') in an experience-dependent manner. Arrows represent information flow via synaptic activity. Engram activity enables the perception itself as well as the creation of a mental image of the butterfly during memory recall (callout symbols on the right).

stage of perception that takes place while you look at the butterfly. Processing of visual information begins in your retina (for simplification represented by an eye), continues in the thalamus (a relay station for incoming sensory input), and the signal then reaches the visual cortex and other specialized cortices (all represented here by pictures of individual pyramidal neurons) before it ultimately arrives at its final destination in associative cortices, such as the prefrontal cortex. The anatomical location of representation is not important for this consideration. Information is typically encoded by populations of neurons, the engrams, which are composed of synaptically connected neurons. During learning, all synapses that were activated along the path of signal processing and representation are 'in use,' and their synaptic weights change as a result of seeing the butterfly. Critically, though, synapses between engram cells that ultimately encode the concept of the butterfly are strengthened so that the same ensemble can be reactivated. This is true even though the cascade of neurons that were originally active upon encountering the butterfly are now inactive and do not drive activity in the memory encoding ensemble, which now becomes a memory engram.

Long-term depression (LTD) also takes place and may result in the reversal of LTP and possibly memory extinction (forgetting). This is a somewhat simplified view as LTD at some synapses within the same circuit might critically support a behavior that depends on LTP at other synapses. However, it is fair to say that memory storage in the adult brain primarily rests on synaptic *potentiation*. This notion is intuitively true as the encoding of new information requires the addition of representation, that is, something 'new' can only be learned by added synapses or added synaptic weight. There are exceptions to this pattern, though. In the cerebellum, a brain area that is primarily involved in motor control, so-called Purkinje cells provide the sole output of the cerebellar cortex. As we will discuss later in this book, these neurons are inhibitory. Therefore, *depression* at their excitatory synaptic inputs enables disinhibition of their target neurons in the cerebellar nuclei. This is why the cerebellum is the only brain area where LTD is considered the dominant type of synaptic plasticity. Exceptions notwithstanding, the notion holds that new information is learned through synaptic potentiation, particularly in the cerebral cortex. The most important conclusion from these considerations is that once information reaches the brain, it is processed and consolidated at the synapses.

Brain plasticity has become a topic of interest to a great number of specialists in disciplines such as neurobiology, psychology, psychiatry, and computer science. At the same time, this topic should not be exclusively accessible to academics. Its implications are too far reaching for understanding normal development and brain developmental disorders, such as autism; for adult learning; and for cognitive decline during aging. None of these complex topics can truly be comprehended without at least a basic understanding of synapses and their involvement in learning. This book has been inspired by introductory lectures that I give at the University of Chicago. The content *per se* has not been overly simplified, but I have tried to make it accessible to an audience without a background in the subject by introducing concepts and terms where required.

# 1 Peter Huttenlocher and the Discovery of Synaptic Pruning

The 1960s and 1970s marked the beginning of an era of research on synapse physiology and synaptic changes underlying brain plasticity. Hallmark studies such as the first publications by David Hubel and Torsten Wiesel on visual deprivation in kittens (Wiesel and Hubel, 1963a, 1963b) as well as the discovery of LTP by Tim Bliss and Terje Lomo (Bliss and Lomo, 1973) are highlights of this period.

In their classic monocular deprivation studies, Hubel and Wiesel demonstrated that the plasticity of neuronal connections and responsiveness is remarkably high during a restricted period of postnatal development.[1] If one eye of a kitten is deprived of visual input during this period, neurons become less responsive to input from that eye, which instead is heavily shifted toward input from the intact eye (Wiesel and Hubel, 1963a, 1963b). Switching the deprivation pattern so that the previously intact eye is deprived of visual input while the previously deprived eye now receives it once more reverses responsiveness, provided the switch happens within the same restricted 'critical period'. These groundbreaking studies will be discussed in depth in Chapter II. Particularly inspired by the Hubel and Wiesel findings, Peter Huttenlocher (Figure 1.1) began his work on the development of synaptic connectivity in the mid-1970s.

Huttenlocher started his academic career as an Assistant Professor in Pediatric Neurology at Harvard University (1964–1966), followed

---

[1] This developmental phase is called the 'critical period.' In cats, it lasts until about 3 months after birth.

9

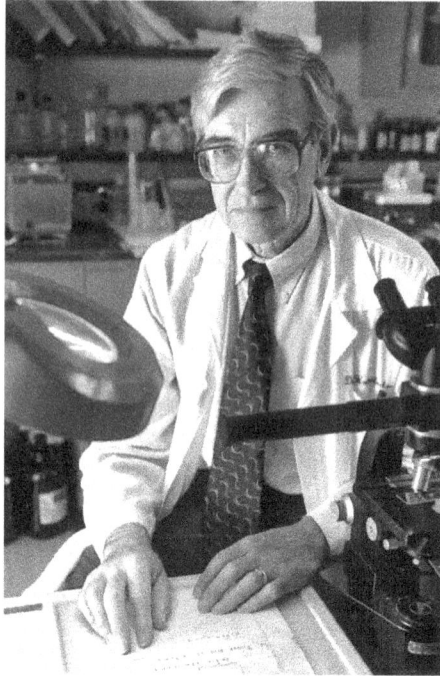

**Figure 1.1:** Peter Huttenlocher (1931–2013). Photograph courtesy of the University of Chicago Photographic Archive, Special Collections Research Center, University of Chicago Library.

by 8 years at Yale University. In 1974, he became a Professor at the University of Chicago. At this time, the full impact of the studies by Hubel and Wiesel had been grasped by the research community, and it was an established fact that brain connectivity was highly plastic during a defined period of early postnatal development. The term 'use it or lose it' became an appropriate catch phrase in communications among neuroscientists of the day. According to personal accounts,[2] Peter Huttenlocher was a quiet and reticent man, certainly not prone to the short-lived fashions that all too often determine the course of scientific research. However, the emerging field of synapse and brain plasticity had a palpable influence on his own work. He developed an interest in the study of synaptogenesis, the formation of synapses

---

[2] See Christopher A. Walsh's orbituary for Peter Huttenlocher in *Nature* (502, p. 172, 2013)

between neurons, as well as synapse elimination during human infancy. Initially, his interest was directed toward abnormal brain development in mental retardation, but soon he developed a fascination with the maturation of the normal healthy brain. Famously, he noted, 'Paradoxically, in our early studies, the findings in the normal population were more interesting than the abnormal population.'[3]

As a pediatric neurologist, Huttenlocher was particularly drawn to the study of human brains, rather than choosing an experimental approach using laboratory animals, as Hubel and Wiesel had done before him. For his first study on synaptic density in the human cortex, which he performed at the University of Chicago, he collected 21 cadaver brains from people who had no intellectual disabilities (covering a typical human lifespan, including the brain of a newborn and the brain of a 90-year-old man). He removed small blocks of their frontal cortices during autopsy and prepared them for electron microscopic analysis using a method that selectively stained synapses.[4] In this seminal study, he showed that at birth, synaptic density was as high as in the adult brain. However, he found that, unexpectedly, synaptic density increases during infancy, peaks between the ages of 1 and 2, and then declines to adult levels between years 2 and 16 (Huttenlocher, 1979). The peak level of synaptic density during early childhood was about 50% higher than that in adult brains. This decline in synaptic density, which begins at a relatively early developmental stage and persists into adulthood, had previously been described in the rat and cat brain (Aghajanian and Bloom, 1967; Cragg, 1975). However, it was Huttenlocher who was credited with the discovery of synaptic pruning, because he was the first to interpret it as an active, competitive elimination process. This first study on its own, though, was not sufficient to reach that conclusion. It was possible that the observed decrease in synaptic density was a result of cortical expansion. A conceivable, although unlikely, scenario was that the increase in cortical volume was largely because of a

---

[3] Reported in a statement from the University of Chicago (J. Easton, *UChicago News*, 2013)

[4] The phosphotungstic acid method of Bloom and Aghajanian (*J. Ultrastruct. Res.* 22, 1968)

strong increase in glial or neuropil components, resulting in a decrease of neuronal—and with it, synaptic—density. Volume meas- urements are not easily made in the frontal cortex because of a lack of clear boundaries to neighboring cortical areas. For this reason, Huttenlocher repeated his study in the human visual cortex (area 17), where the boundaries are distinct. This work demonstrated that the volume of area 17 reached its maximum at about 4 years of age, at which time synaptic density continued to decline. This finding sug- gested that the drop in synaptic density is an elimination process and not the mere result of changes in cortical volume (Huttenlocher *et al.*, 1982). Figure 1.2 shows Peter Huttenlocher's findings on synaptic density and the total number of synapses in area 17, plotted as a function of age (Huttenlocher, 1990). Two phases in synaptic devel- opment stand out: first a sharp increase in synapse number and den- sity that peaks at about 8 months of age, followed by a slower decline

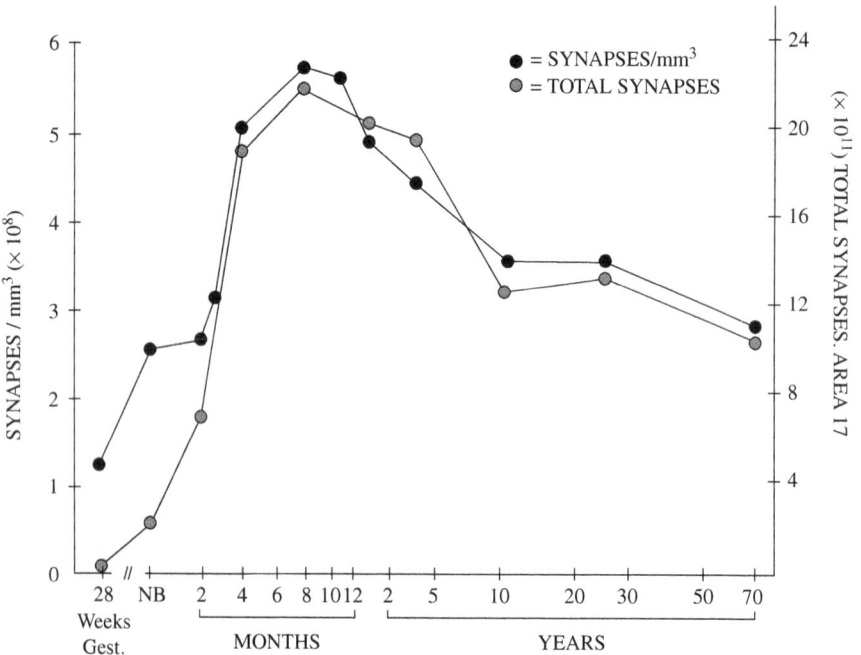

**Figure 1.2:** Synaptic density and total synapses in visual cortex as a function of age.[5]

[5] Reproduced from Huttenlocher (1990).

that persists until a time point between 5 and 10 years of age, and levels off after that. As previously observed in the frontal cortex, the peak level of synaptic density was close to 50% higher than that measured in adult brains. In a later study, Huttenlocher also examined synaptic density profiles in the human auditory cortex and again found that a sharp rise in synaptic density was followed by synapse elimination. In this comparative study, he noted differences in timing: the peak in synaptic density was reached in the auditory and visual cortices at about 3 and 8 months, respectively, whereas in the prefrontal cortex it was reached at 1–3 years of age. This demonstrates that cortical maturation differs depending on the specialization of the respective cortical area (Huttenlocher and Dabholkar, 1997).

The same general pruning phenomenon was also observed—subsequent to Huttenlocher's initial findings—in nonhuman primates (Rakic *et al.*, 1986) and in mice (Zuo *et al.*, 2005; Bian *et al.*, 2015), animal species that are currently used for the study of brain developmental disorders, where pruning deficits might well play a role. The significance of Huttenlocher's findings can only be fully appreciated within the context of the studies by Hubel and Wiesel. Their discovery of a critical period in brain development, during which synapses either weaken and disconnect or stabilize, demonstrated that competition between synaptic inputs is a guiding principle in the formation of functional brain circuits. Huttenlocher's work showed that, in fact, synapse elimination dominates during late childhood and that this phenomenon has an unexpected magnitude: up to 50% of all synapses disappear.[6] Therefore, an optimized brain architecture is 'carved out' of an overabundance of synaptic connections that is typical of the early postnatal brain (Figure 1.3).

Huttenlocher's observation that synaptic pruning occurred during a restricted time window that ended before age ten aligned well with the critical period that was observed in the visual system of kittens by Hubel and Wiesel. Accepting critical periods of synaptic re-organization as a general motif in brain development, he became a champion of early

---

[6]Note that this is a net change: there is an ongoing turnover of synapses, during which individual synapses are replaced by new ones. Many synapses undergo potentiation. Overall, there is, however, a reduction in synapse number.

**Synaptic Pruning**

——————→

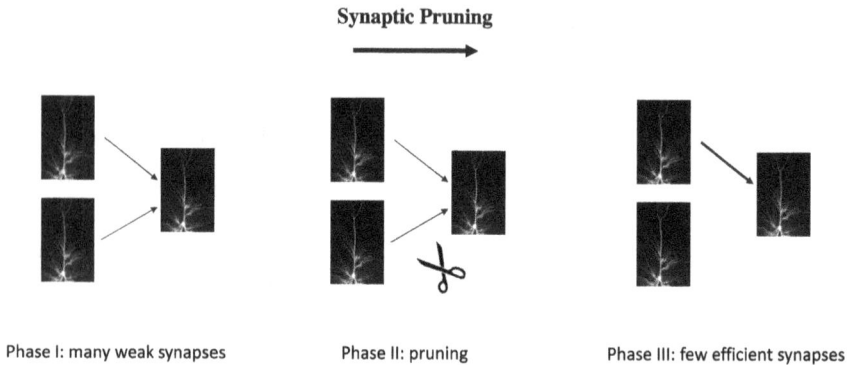

Phase I: many weak synapses        Phase II: pruning        Phase III: few efficient synapses

**Figure 1.3:** Synaptic pruning is a competitive selection process that eliminates all inefficient synaptic connections, leaving a lower number of 'winner' synapses intact.

child education—for example, in learning to play music instruments or acquiring foreign languages.[7] His work had a large impact on the development of educational approaches by psychologists and pedagogues. In his interpretation of synaptic overabundance during childhood, Peter Huttenlocher followed the argument of Jean-Pierre Changeux that the early synaptic contacts are random and redundant (Changeux and Danchin, 1976). However, the selection of 'winner' synapses is certainly not a random process. As I will discuss later in this book (Chapters II and III), coding accuracy is a crucial factor, meaning that the activity of synapses that become 'winners' must be scaled against meaningful activity patterns in the surrounding population of neurons.

Describing the significance of Huttenlocher's work, his colleague, the Nobel Prize winner Eric Kandel (Columbia University) once said: 'There was the suggestion, based on animal studies, that humans might assemble these connections between neurons at birth. But no one was thinking about young children subsequently losing those connections. We now know how absolutely crucial synaptic pruning is to mental development and that defects in this system can lead to severe cognitive deficits.'[3] A prominent example is that synaptic pruning is impaired in autism spectrum disorders (ASDs), where

_____

[7] Huttenlocher explicitly expressed this view in an editorial letter published in *Nature Neuroscience* (Vol. 6, 2003).

synapse density[8] remains elevated well beyond the normal pruning period (Hutsler and Zhang, 2010; Tang *et al.*, 2014). In Chapter V, we will discuss synapse pruning deficits in ASD in more detail. Huttenlocher's work has brought to light the magnitude of plastic changes in the developing human brain. In contrast to most animal species, human development and adolescence span many years. More than any studies on experimental animals, Huttenlocher's work has demonstrated that adaptive plasticity does not have to be restricted to a short episode after birth. In fact, its prolonged nature in the human brain suggests that extensive brain maturation consti-tutes an essential component of long-lasting developmental periods, and perhaps even a reason for such long duration. In the following chapters, I will outline what synaptic plasticity means for brain matu-ration and adaptation.

# References

Aghajanian, G.K., and Bloom, F.E. (1967). The formation of synaptic junc-tions in developing rat brain: a quantitative electron microscopic study. *Brain Res.* 6, 716–727.

Bian, W.J., Miao, W.Y., He, S.J., Qiu, Z., and Yu, X. (2015). Coordinated spine pruning and maturation mediated by inter-spine competition for cadherin/catenin complexes. *Cell* 162, 808–822.

Bliss, T.V.P., and Lomo, T. (1973). Long-lasting potentiation of synaptic trans-mission in the dentate area of the anaesthetized rabbit following stimu-lation of the perforant path. *J. Physiol.* 232, 331–356.

Changeux, J.P., and Danchin, A. (1976). Selective stabilization of develop-ing synapses as a mechanism for the specification of neuronal networks. *Nature* 264, 705–712.

Cragg, B.G. (1975). The development of synapses in the visual system of the cat. *J. Comp. Neurol.* 160, 147–166.

Hutsler, J.J., and Zhang, H. (2010). Increased dendritic spine densities on cortical projection neurons in autism spectrum disorders. *Brain Res.* 1309, 83–94.

Huttenlocher, P.R. (1979). Synaptic density in human frontal cortex—developmental changes and effects of aging. *Brain Res.* 163, 195–205.

---

[8]Measured as the density of dendritic spines. Spines are small protrusions located on neuronal dendrites that form the contact sites for excitatory synapses.

Huttenlocher, P.R. (1990). Morphometric study of human cerebral cortex development. *Neuropsychologia* 28, 517–527.

Huttenlocher, P.R., and Dabholkar, A.S. (1997). Regional differences in synaptogenesis in human cerebral cortex. *J. Comp. Neurol.* 387, 167–178.

Huttenlocher, P.R., de Courten, C., Garey, L.J., and van der Loos, H. (1982). Synaptogenesis in human visual cortex—evidence for synapse elimination during normal development. *Neurosci. Lett.* 33, 247–252.

Rakic, P., Bourgeois, J.P., Eckenhoff, M.F., Zecevic, N., and Goldman-Rakic, P.S. (1986). Concurrent overproduction of synapses in diverse regions of the primate cerebral cortex. *Science* 232, 232–235.

Tang, G., *et al.* (2014). Loss of mTOR-dependent macroautophagy causes autistic-like synaptic pruning deficits. *Neuron* 83, 1131–1143.

Wiesel, T.N., and Hubel, D.H. (1963a). Effects of visual deprivation on morphology and physiology of cells in the cat's lateral geniculate body. *J. Neurophysiol.* 26, 978–993.

Wiesel, T.N., and Hubel, D.H. (1963b). Single-cell responses in striate cortex of kittens deprived of vision in one eye. *J. Neurophysiol.* 26, 1003–1017.

Zuo, Y., Lin, A., Chang, P., and Gan, W.B. (2005). Development of long-term dendritic spine stability in diverse regions of cerebral cortex. *Neuron* 46, 181–189.

# 2 Why Postnatal Brain Plasticity Is Needed: Limitations of Genetic Blueprints

The developmental elimination of surplus synapses is the core component of an extensive reorganization of neural networks during early life. Why does the brain go through the energy-consuming process of generating an excess of synaptic contacts, only to have about half of them subsequently removed? This problem does not equally apply to adult plasticity, where changes in synaptic connectivity are less dramatic. New information that is acquired needs to be processed and stored in adult life. Throughout life, changes in the environment necessitate appropriate 'upgrades' in brain circuits, involving synaptic potentiation as well as depression events. In early brain maturation, however, the need for adaptive plasticity is less obvious. Why is there no blueprint of synaptic connectivity that is genetically determined and only requires upgrades when needed, similar to the adaptive processes that take place in the adult brain?

To be sure, there is a genetic blueprint in the form of a crude connectivity pattern. The molecular machinery involved in the preprogrammed sequence of plasticity events during postnatal development is genetically set as well. This is different in adult plasticity: in the absence of appropriate stimuli and events, synapses and circuits in adult life might reach a point of stability.[1] Why is a 'soft' upgrade not

---

[1] This is obviously a theoretical construct. A state of complete absence of stimuli does not exist.

the best strategy for the developing brain? In short, the answer is that the optimal circuit architecture is not known at birth. It depends, among other factors, on the development of morphological features of the skull, the sensory organs, and other brain circuits, all of which are influenced by 'environmental' factors such as nutrition. While these morphological features take shape, development of the optimal brain architecture must follow. This in an important point to make. Let me develop the argument in steps, based on the findings of Hubel and Wiesel, which I referenced in Chapter 1.

David H. Hubel and Torsten N. Wiesel (Figure 2.1) started what was to become one of the most fruitful collaborations in neuroscience when they met at Johns Hopkins University. Their first study as a team was published in 1959 in *The Journal of Physiology* (Hubel and Wiesel, 1959). They subsequently moved to Harvard University, where they did most of the groundbreaking work that earned them the Nobel Prize in Physiology or Medicine in 1981.

Hubel and Wiesel selected the visual system of the cat for their initial experiments. This choice was motivated by the ability to

**Figure 2.1:** Torsten Wiesel (left) and David Hubel (right) in their laboratory. The photograph is reproduced from Brain and Visual Perception. The Story of a 25-year Collaboration by David Hubel (2004). Copyright: Oxford Publishing Limited. Reproduced with permission of the Licensor through PLSclear.

present sensory stimuli that could easily be generated and controlled—they used a projector as a light source—and by the comparably rich knowledge about this brain system that was available at the time. The processing of visual signals begins in the retina of the eye, where light-sensitive cells, the so-called rods and cones, convert sensory input into electric signals that reach ganglion cells in the retina. The ganglion cell axons from each eye form the optic nerve, which partially crosses the midline at the optic chiasm and reaches the lateral geniculate nucleus (LGN) of the thalamus, a complex of nuclei that constitutes a relay station on the path of these signals to the cortex. Sensory information from the retina first reaches the primary visual cortex (or striate cortex) and is subsequently transferred to other cortical areas. At each synaptic transfer, signal processing and extraction of signal features takes place in a way that has been characterized in detail, and is described elsewhere (e.g. Hubel and Wiesel, 1979).

Hubel and Wiesel performed single-unit recordings from the striate cortex of the cat using tungsten electrodes that were lowered into the tissue perpendicular or at a defined angle to the cortical surface. They measured response properties of neurons in different layers of the striate cortex and observed what they called 'ocular dominance columns'. Within a given column of cortical tissue, perpendicularly passing the six layers of cortex, neurons predominantly responded to input from either the left or the right eye (Hubel and Wiesel, 1959, 1962, 1979). Neurons in the neighboring column responded more strongly to input from the other eye. Layer IV differed from the other layers, as neurons in this layer exclusively responded to input from one eye (the eye that cells in the layers above and below preferentially responded to). In other words, they showed monocular responses. The outcome of the recordings from cats is illustrated in Figure 2.2.

In histological studies of the brains of macaque monkeys, they used Louis Sokoloff's 2-deoxyglucose staining technique, and were able to visualize the ocular dominance columns. Active neurons take up glucose as an energy source. In contrast to glucose, 2-deoxyglucose cannot be fully metabolized and therefore accumulates in cells that have been active for some time. Radioactive

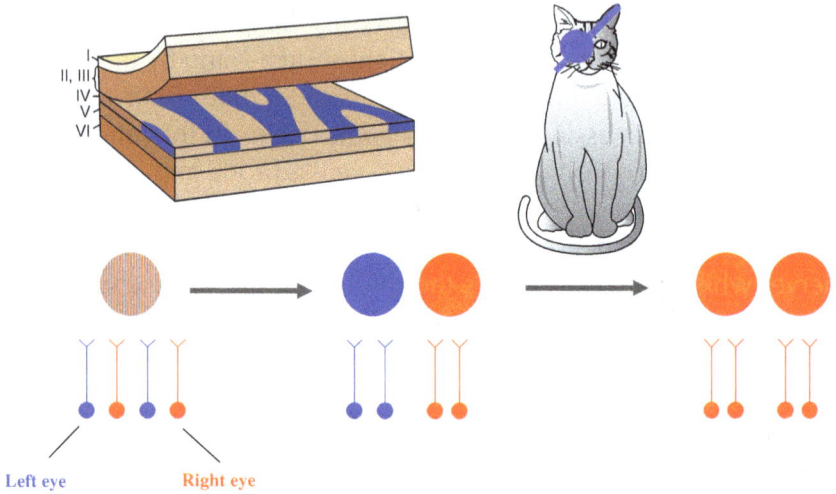

**Figure 2.2:** Plasticity of ocular dominance columns. In an adult cat, neurons in cortical layer IV receive afferents from either the left eye (from the perspective of the observer; blue) or the right eye (orange). Observations in monkeys and ferrets indicate that at birth, a raw ocular dominance pattern might already exist that, however, is likely to differ from the adult pattern in the precision of input separation (left; the circle on top indicates a neuron in an immature column that still shows a level of input overlap—exaggerated here for illustrative purposes). During a subsequent pruning period, input from one of the eyes is eliminated, so that each neuron in layer IV receives afferents from only one eye (middle; the two circles symbolize representative neurons from each column). If input to one eye is lost (illustrated here through a patch to one eye), afferents from the intact eye take over (right; note that this scheme slightly exaggerates the loss of input from the deprived eye: although many afferents are indeed disconnected, some afferents remain in place). This alteration can be reversed upon switching which eye is closed, as long as it happens during the critical period (1–3 months after birth). The same general type of input organization and plasticity takes place in neurons from layers above and below layer IV, but these neurons are binocular, with dominance of input from one eye.

labeling of 2-deoxyglucose with $C^{14}$ leads to a lasting label of such neurons. When light is presented to only one eye for about 45 minutes, the striped activity pattern that reflects ocular dominance can be detected in brain sections that are subsequently prepared. The width of an individual column cycle ('*column cycle*' is a term that is used to describe the left and right input domain together) was found to be 0.5–0.8 mm (LeVay *et al.*, 1980; see also Hubel *et al.*, 1977), whereas in cats the column cycle is slightly wider (1.0–1.1 mm; Löwel and Singer, 1987).

In the so-called monocular deprivation experiments, Hubel and Wiesel studied the plasticity of cortical connectivity. They deprived kittens of visual input to one eye before they would open their eyes (about seven to ten days after birth). This was done by suturing the upper and lower eyelids together, a technique that offered the advantage that the suture could easily be removed at any time during the cat's development (eye patches provide an alternative method in some species). When performing the same type of single-unit recordings as in their previous studies, but now one to four months after eye closure, they found that the ocular dominance pattern had dramatically changed. Essentially all neurons in striate cortex now responded to visual input from the intact eye, but none responded to visual input from the deprived eye, whose lids were now separated for the recording (Figure 2.2; Wiesel and Hubel, 1963). Surprisingly, Hubel and Wiesel found that this plasticity was restricted to a critical period[2] in postnatal development: susceptibility to monocular deprivation began near the beginning of the fourth week (when kittens start using their eyes) and strong deprivation effects were seen up to two months after birth. Then, susceptibility declined and the sensitive period ended after three months. Four months after birth, visual deprivation no longer had an effect on response patterns (Hubel and Wiesel, 1970). *Reverse suturing*, that is, the opening of the previously deprived eye and closure of the previously open eye, completely switches the ocular dominance pattern as long as the procedure is performed within the critical period (Blakemore and Van Sluyters, 1974; Movshon, 1976).

Importantly, pruning processes do not always start from a state of complete input overlap with no structure to the input pattern.

---

[2] Hubel and Wiesel originally used the term 'sensitive period'.

This appears to be true for the ocular dominance pattern as well. Early studies have reported overlap between left-eye and right-eye inputs to striate cortex in monkeys and cats (Rakic, 1976; LeVay *et al.*, 1978). However, later work—using improved staining techniques—revealed the existence of a columnar structure in newborn monkeys, prior to visual experience (Horton and Hocking, 1996; for similar findings in ferrets, see Crowley and Katz, 2000). This early, experience-independent pattern forms before the opening of the eyes, guided by wave-like activity in the retina (Stellwagen and Shatz, 2002). These preformed patterns become functional columns during the days to follow, now guided by the activity-dependent corrective processes that Hubel and Wiesel described. Circuit optimization is still the organizing principle, assisting in maintaining and sharpening the stripe pattern while the cortex volume increases during development, and providing an efficient mechanism to eliminate faulty connections.

Hubel and Wiesel were experts in the neuroscience of vision, but they were not synapse physiologists *per se*. The primary motivation for their work was to delineate the functional architecture of the visual system and the representation of visual stimuli in cortex. However, they noted, 'Columnar development therefore seems likely to involve breakage and re-formation of functional synapses' (LeVay *et al.*, 1980), demonstrating that they were aware of the synaptic nature of the restructuring processes that they observed, and that they appreciated the relevance of their findings for early brain development. However, as their experiments focused on the processing of visual information, rather than the fate of individual synapses, they never directly addressed the question of whether the total number of synapses changed when the 'breakage' happened, and thus did not discuss their findings in the context of pruning. It was ultimately Peter Huttenlocher who demonstrated that dramatic changes in the number of synaptic contacts are a hallmark of early brain development (Huttenlocher, 1979). And yet, Hubel and Wiesel contributed immensely to our understanding of developmental pruning by conducting the first studies of brain plasticity and the physiological processes that are involved.[3] A later generation of neurophysiologists characterized the cellular events underlying the plasticity of monocular deprivation with a greater focus on the synapse itself (Singer, 1995; Bear and Rittenhouse, 1999). This step ultimately allowed them to focus on the molecular machinery involved, along with the comparison to plasticity mechanisms in the adult brain (see Piochon *et al.*, 2016). I will discuss this complex topic later in the book (Chapter IV).

Other impressive examples of changes in brain connectivity as a result of sensory deprivation have been reported that involve the interplay of different sensory modalities. In the short-tailed opossum, loss of visual input results in an expansion of auditory and somatosensory cortices at the expense of the visual cortex (Kahn and Krubitzer, 2002). Similarly, studies of mice that are deaf at birth have shown that the auditory cortex is encroached by the visual and somatosensory cortex (Hunt *et al.*, 2006). These examples, together with the studies by Hubel and Wiesel, show that functional assignment in cortical development depends on sensory experience, which cannot always be predicted to the degree necessary for neural circuit formation.

Let us return to the question that was posed at the beginning of this chapter. Why is it not enough to determine a synaptic connectivity map via a genetic blueprint? The reason is that environmental factors alter what the 'optimal' genetic blueprint is. These include metabolic factors that influence growth and, ultimately, the size of body structures, as well as 'epigenetic' factors experienced during development, which cause changes in gene expression without altering the DNA sequence. (The term 'epigenetic' was coined by Conrad Waddington, a British developmental biologist, in 1942.) To consider such environmental factors, we can return to the example of the ocular dominance columns described by Hubel and Wiesel. With the

---

[3] Excellent additional model synapses for the study of pruning are the neuromuscular junction and the climbing fiber input to Purkinje cells in the cerebellum (reviewed in Piochon *et al.*, 2016).

exception of monocular layer IV, all cortical layers have neurons that respond binocularly, that is, they receive input from both eyes. The 'dominance' arises from the fact that these neurons respond more strongly to input from one eye. Nevertheless, they can also be activated by input from the nondominant eye. These neurons receive their visual inputs from corresponding areas of the two retinas, which are activated by light from the exact same points in the visual field. Environmental factors alter the locations of these corresponding retinal areas as well as the morphology of the cortex.

The availability of the vitamin folic acid, for example, regulates the rate of neurogenesis and apoptosis[4] in the developing fetal brain, and with it the size of the cortical surface area, the so-called 'cortical sheet' (Craciunescu et al., 2004, 2010). This affects cortex dimensions and architecture. Dietary factors, such as the intake of fat, vitamins B1, B2, and C as well as iron and cholesterol, impact the shape of the eyeball (Edwards, 1996). Dietary calcium intake determines the rate of bone growth (e.g. Carttar et al., 1950), likely contributing to the shaping of craniofacial morphology, which is also affected by diet and the resulting chewing behavior[5] (He, 2004; Koyabu and Endo, 2009). These findings suggest that diet constitutes an important environmental factor that will impact the shape of the eyeball and its surrounding skull structure; the eye socket, or 'orbit'; as well as the morphology of target areas of the retino-thalamo-cortical projections in the striate cortex. [For a review on environmental factors in cortical organization, see Krubitzer and Dooley (2013).]

Imagine if connectivity were exclusively determined by a genetic blueprint. There is no efficient anticipation of how morphological parameters, such as the length of the eyeball, will develop. It is therefore impossible to genetically predetermine optimal synaptic connectivity that would relay input from exactly corresponding retinal areas to their cortical target zones. These considerations make it plausible that experience-dependent pruning, starting from an overabundance of synaptic connections, is indeed the most efficient way

---

[4] Apoptosis means programmed cell death.
[5] The scientific term for chewing is 'mastication'.

to develop brain circuits during early postnatal life. In addition to such metabolic factors, epigenetic influences control gene transcription through processes such as the phosphorylation and acetylation of histones, which are DNA-binding proteins (Putignano *et al.*, 2007), as well as DNA methylation (Sweatt, 2019). As a result, it is conceivable that the epigenetic regulation of any gene that determines the location, relative size, connectivity, and patterning of cortical areas, such as *Fgf8*, the gene coding for fibroblast growth factor 8 (growth factors are substances that promote cellular growth and/or proliferation; Fukuchi-Shimogori and Grove, 2001; Assimacopoulos *et al.*, 2012), might alter cortical circuit architectures in a way that necessitates postnatal rewiring and plasticity.

The duration of critical periods for cortical development depends on the species, type of cortex, and function tested. The more complex the cortical structure, the longer the critical period. Hubel and Wiesel determined a critical period of about three months for development of the visual cortex in cats (Wiesel and Hubel, 1963). In monkeys, the critical period for cortical development is in the range of one to two years, whereas in humans it lasts about five to six years (Crawford *et al.*, 1993). In children, the reduction of cortical plasticity after closure of this time frame makes it necessary to correct for ophthalmologic problems, such as strabismus (a misalignment of the two eyes resulting from a lack of appropriate eye muscle control) or cataracts (opaque lens) as early as possible, but certainly before the child reaches school age (reviewed in Huttenlocher, 2002). The criticality of such corrections also arises from the observation that it is not only synapses that are affected by critical period plasticity. During this time, axonal projections of 'loser inputs' are rebuilt and degenerate.

Unsurprisingly, the best examples of developmental brain plasticity come from studies of sensory systems. Sensory input can be readily modified and systematically varied, and responses to stimuli can be recorded from well-defined cortical areas. Similar physiological studies are not available for regions of cortex that are not primary sensory cortical areas, such as motor cortex, associative cortices, or prefrontal cortex. Nevertheless, pruning has been anatomically demonstrated in these and other noncortical areas as well. It is therefore safe to assume that the general motif of pruning by competitive plasticity takes place throughout the brain and follows the principles established by David Hubel and Torsten Wiesel. Likewise, we know that the phenomenon of developmental critical periods extends throughout the cerebral cortex and likely the entire brain. This is why, as mentioned before, Huttenlocher, later on in his academic career, emphasized the relevance of critical periods for positive educational influences on brain development in children, such as in language acquisition and music training (e.g. Huttenlocher, 2002, 2003).

In closing this chapter on the early studies of developmental brain plasticity by Hubel and Wiesel, I will share an observation that I made while writing. Hubel and Wiesel performed most of their essential work alone, without additional help from students or postdoctoral assistants. They published their most important work in journals such as *The Journal of Physiology* and *Journal of Neurophysiology*, which focus on science as opposed to flashy headlines, even though more glamorous journals, such as *Science* and *Nature,* existed. They published their reputation-defining work in a small number of papers, which, at least for those publications, amounted to a relatively low h-index.[6] This was true of Peter Huttenlocher as well. Yet, all three of them were giants in their fields and left a mark on neuroscience like few others. By current numerical standards of success, which favor large laboratories, glossy journals, and high publication volumes, Hubel, Wiesel, and Huttenlocher would appear, on paper, as moderately successful. Under the pressures of grant applications and review, which rely heavily on fast assessment of productivity and success, something has been lost: the appreciation of true value in scientific research.

---

[6]The h-index is a metric for the productivity of a scientist, which is defined as the number of *h* papers that have been cited at least *h* times. For example, a scientist who published 21 papers, of which number 18 (ranked by number of citations) has been cited 18 times, whereas number 19 has been published less than 19 times, has an h-index of 18.

# References

Assimacopoulos, S., Kao, T., Issa, N.P., and Grove, E.A. (2012). Fibroblast growth factor 8 organizes the neocortical area map and regulates sensory map topography. *J. Neurosci.* 32, 7191–7201.

Bear, M.F., and Rittenhouse, C.D. (1999). Molecular basis for induction of ocular dominance plasticity. *J. Neurobiol.* 41, 83–91.

Blakemore, C., and Van Sluyters, R.C. (1974). Reversal of the physiological effects of monocular deprivation in kittens: further evidence for a sensitive period. *J. Physiol.* 237, 195–216.

Carttar, M.S., McLean, F.C., and Urist, M.R. (1950). The effect of the calcium and phosphorus content of the diet upon formation and structure of bone. *Am. J. Pathol.* 26, 307–331.

Craciunescu, C.N., Brown, E.C., Mar, M.H., Albright, C.D., Nadeau, M.R., and Zeisel, S.H. (2004). Folic acid deficiency during late gestation decreases progenitor cell proliferation and increases apoptosis in fetal mouse brain. *J. Nutr.* 134, 162–166.

Craciunescu, C.N., Johnson, A.R., and Zeisel, S.H. (2010). Dietary choline reverses some, but not all, effects of folate deficiency on neurogenesis and apoptosis in fetal mouse brain. *J. Nutr.* 140, 1162–1166.

Crawford, M.L., Harwerth, R.S., Smith, E.L., and von Noorden, G.K. (1993). Keeping an eye on the brain: the role of visual experience in monkeys and children. *J. Gen. Psychol.* 120, 7–19.

Crowley, J.C., and Katz, L.C. (2000). Early development of ocular dominance columns. *Science* 290, 1321–1324.

Edwards, M.H. (1996). Do variations in normal nutrition play a role in the development of myopia? *Optom. Vis. Sci.* 73, 638–643.

Fukuchi-Shimogori, T., and Grove, E.A. (2001). Neocortex patterning by the secreted signaling molecule FGF8. *Science* 294, 1071–1074.

He, T. (2004). Craniofacial morphology and growth in ferret: effects from alteration of masticatory function. *Swed. Dent. J. Suppl.* 165, 1–72.

Horton, J.C., and Hocking, D.R. (1996). An adult-like pattern of ocular dominance columns in striate cortex of newborn monkeys prior to visual experience. *J. Neurosci.* 16, 1791–1807.

Hubel, D.H., and Wiesel, T.N. (1959). Receptive fields of single neurones in the cat's striate cortex. *J. Physiol.* 148, 574–591.

Hubel, D.H., and Wiesel, T.N. (1962). Receptive fields, binocular interaction and functional architecture in the cat's visual cortex. *J. Physiol.* 160, 106–154.

Hubel, D.H., and Wiesel, T.N. (1963). Receptive fields of cells in striate cortex of very young, visually inexperienced kittens. *J. Neurophysiol.* 26, 994–1002.

Hubel, D.H., and Wiesel, T.N. (1970). The period of susceptibility to the physiological effects of unilateral eye closure in kittens. *J. Physiol.* 206, 419–436.

Hubel, D.H., and Wiesel, T.N. (1979). Brain mechanisms of vision. *Sci. Am.* 241, 150–162.

Hubel, D.H., Wiesel, T.N., and Stryker, M.P. (1977). Orientation columns in macaque monkey visual cortex demonstrated by the 2-deoxyglucose autoradiographic technique. *Nature* 269, 328–330.

Hunt, D.L., Yamoah, E.N., and Krubitzer, L. (2006). Multisensory plasticity in congenitally deaf mice: how are cortical areas functionally specified? *Neuroscience* 139, 1507–1524.

Huttenlocher, P.R. (1979). Synaptic density in human frontal cortex — developmental changes and effects of aging. *Brain Res.* 163, 195–205.

Huttenlocher, P.R. (2002). *Neural Plasticity: The Effects of Environment on the Development of the Cerebral Cortex* (Cambridge, MA, Harvard University Press).

Huttenlocher, P.R. (2003). Basic neuroscience research has important implications for child development. *Nat. Neurosci.* 6, 541.

Kahn, D.M., and Krubitzer, L. (2002). Massive cross-modal cortical plasticity and the emergence of a new cortical area in developmentally blind mammals. *Proc. Natl. Acad. Sci. USA* 99, 11429–11434.

Koyabu, D.B., and Endo, H. (2009). Craniofacial variation and dietary adaptations of African colobines. *J. Hum. Evol.* 56, 525–536.

Krubitzer, L., and Dooley, J.C. (2013). Cortical plasticity within and across lifetimes: how can development inform us about phenotypic transformations? *Front. Hum. Neurosci.* 7, 620.

LeVay, S., Stryker, M.P., and Shatz, C.J. (1978). Ocular dominance columns and their development in layer IV of the cat's visual cortex: a quantitative study. *J. Comp. Neurol.* 179, 223–244.

LeVay, S., Wiesel, T.N., and Hubel, D.H. (1980). The development of ocular dominance columns in normal and visually deprived monkeys. *J. Comp. Neurol.* 191, 1–51.

Löwel, S., and Singer, W. (1987). The pattern of ocular dominance columns in flat-mounts of the cat visual cortex. *Exp. Brain Res.* 68, 661–666.

Movshon, J.A. (1976). Reversal of the physiological effects of monocular deprivation in the kitten's visual cortex. *J. Physiol.* 261, 125–174.

Piochon, C., Kano, M., and Hansel, C. (2016). LTD-like molecular pathways in developmental synaptic pruning. *Nat. Neurosci.* 19, 1299–1310.

Putignano, E., Lonetti, G., Cancedda, L., Ratto, G., Costa, M., Maffei, L., and Pizzorusso, T. (2007). Developmental downregulation of histone post-translational modifications regulates visual cortical plasticity. *Neuron* 53, 747–759.

Rakic, P. (1976). Prenatal genesis of connections subserving ocular dominance in the rhesus monkey. *Nature* 261, 467–471.

Singer, W. (1995). Development and plasticity of cortical processing architectures. *Science* 270, 758–764.

Stellwagen, D., and Shatz, C.J. (2002). An instructive role for retinal waves in the development of retinogeniculate connectivity. *Neuron* 33, 357–367.

Sweatt, J.D. (2019). The epigenetic basis of individuality. *Curr. Opin. Behav. Sci.* 25, 51–56.

Waddington, C.H. (1942). The epigenotype. *Endeavour* 1, 18–20.

Wiesel, T.N., and Hubel, D.H. (1963). Single-cell responses in striate cortex of kittens deprived of vision in one eye. *J. Neurophysiol.* 26, 1003–1017.

# 3  What Goes Up Must Come Down: Synaptic Potentiation and Depression

In the adult brain, plasticity is no longer about the massive reorganization of connectivity that happens during the pruning phase in childhood. Instead, synaptic weights are adjusted in more subtle ways. Even though the changes are less dramatic during adulthood than during development, these changes constitute responses to unique and personal experiences that become 'engraphed' in our brain circuits. This type of plasticity is a biological phenomenon, not a metaphysical one, but it provides the basis for the absolute and unrestricted individuality that we all possess. The cellular machinery that translates experiences into synaptic weight changes is discussed in this chapter.

Synaptic potentiation constitutes the default mode of synaptic memory formation because new information is added and stored via the strengthening of those synaptic inputs that convey it. This concept is illustrated in Figure 0.1. Synapses involved in the perception of a visual object (here, a butterfly) potentiate their input weights. If this happens at synapses connecting 'butterfly engram' cells, the engram becomes more stable, and recall of the butterfly memory is facilitated. The adult brain uses this mechanism to store memories across the different brain regions.[1] This is not to say that the developing brain cannot use synaptic potentiation toward the same goals. In fact, the

---

[1] There is no one 'learning center'. All brain areas store memories related to the task that they perform.

(a)

(b)                                        (c)

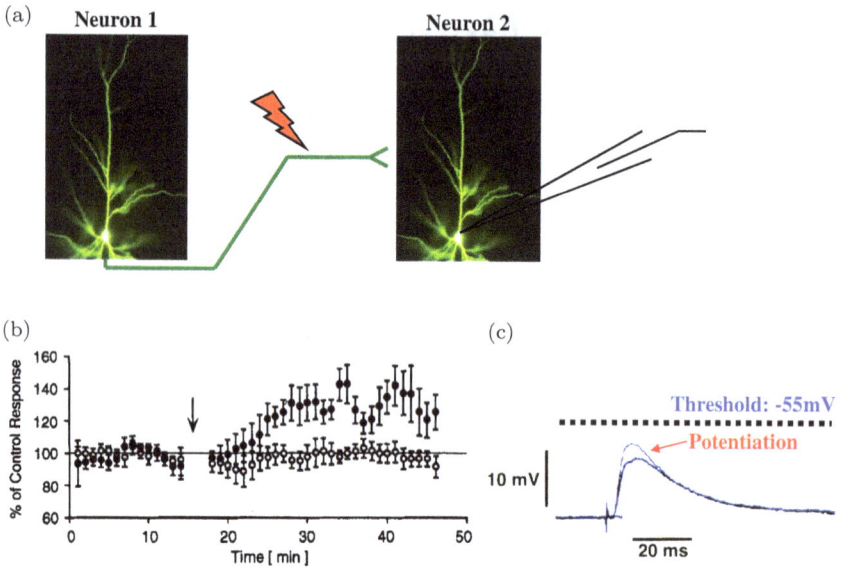

**Figure 3.1:** Synaptic long-term potentiation (LTP). (a) The axon from a neuron A to a neuron B can be electrically stimulated, while responses to synaptic activation are recorded in neuron B. (b) Typical example of an LTP experiment in slices from the rat visual cortex. Plotted are the amplitudes of the excitatory postsynaptic potential (EPSP) over time, normalized to the baseline before tetanization (arrow). Synaptic responses are recorded from layer 2/3 pyramidal neurons in the primary visual cortex (V1). For LTP induction, synaptic afferents are tetanized repeatedly at 50Hz. The black dots show LTP in a tetanized pathway; the white dots show responses in a control pathway that was not tetanized (n=7). (c) The enhanced ('potentiated') EPSP brings the response amplitude closer to the spike threshold (which in this example illustration is set to −55 mV). Panels (b) and (c) are adapted from Hansel *et al.* (1997).

large-scale synaptic reorganization observed after depriving an animal from a specific sensory input demonstrates the need for potentiation and stabilization mechanisms quite well. In Chapter II, we discussed impressive examples of synaptic map plasticity that were studied by Leah Krubitzer at UC Davis. Krubitzer showed that auditory and somatosensory cortices expand at the expense of the visual cortex upon visual deprivation (Kahn and Krubitzer, 2002), whereas deafness leads to an expansion of the visual and somatosensory cortices at the expense of the auditory cortex (Hunt *et al.*, 2006). Plucking of whiskers in rodents similarly initiates an alteration of the representation of the remaining whisker(s) in the somatosensory cortex. Inputs representing the remaining whiskers expand into deprived cortical columns, which, in the naïve state, predominantly receive their input from one specific whisker[2] (Feldman and Brecht, 2005). Synaptic reorganization involves both weakening/disconnection (of synapses conveying information about the deprived input) and strengthening/stabilization (of synapses conveying information about the non-deprived inputs). The expansion of cortical input territory thus depends upon synaptic potentiation events. The role of potentiation in these examples of circuit reorganization in the developing brain and likewise in the memorization of new signals in the adult brain (which this chapter focuses on) is to enable responsiveness to meaningful input signals. From a cell physiological perspective, this is achieved when LTP brings the EPSP amplitude—which represents the strength of the response to a particular signal—closer to spike threshold. This process is illustrated in Figure 3.1.

The figure also highlights two key features of LTP: *activity-dependence* and *input specificity*. 'Activity-dependence' describes the fact that tetanization (enhanced synaptic activation) is required to initiate plasticity changes. In the intact animal, the corresponding term is *experience-dependence*. In Figure 3.1(b), this effect can be seen in the timing of potentiation, whose onset is locked to the tetanization event (arrow). 'Input specificity' means that only the tetanized input potentiates, while an unstimulated control input remains at its original amplitude level. Other, typically weaker, tetanization protocols initiate LTD instead, which can be triggered from a naïve baseline,[3] but also from a potentiated state. In the adult brain, bidirectional plasticity is typically less dramatic than in the

---

[2]In layer 4, these input clusters form the so-called 'barrels', hence this area of somatosensory cortex is called 'barrel cortex'.

[3]The 'naïve baseline' is a construct that arises from experimental considerations. It describes the EPSP amplitude before an experimental intervention. Positive or negative changes relative to this baseline can then be described as potentiation or depression. A somewhat more useful description is that of a linear scale of input weights, which can be assumed by synapses. Within this conceptual framework, 'potentiation' and 'depression' refer to the immediately preceding state of the synapse (the history of synaptic weights), not to a theoretical naïve baseline.

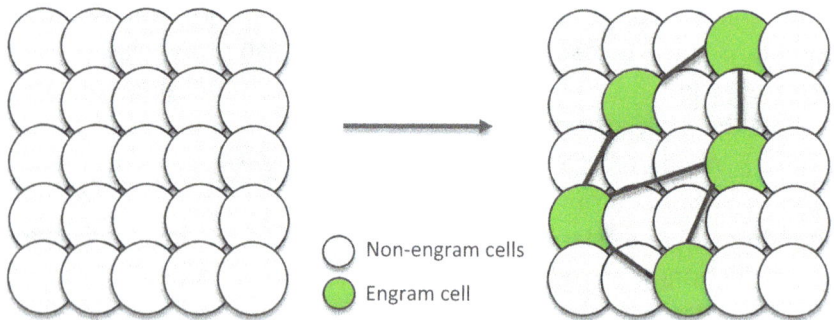

**Figure 3.2:** A neural circuit before (left) and after learning (right). The engram consists of a group of connected neurons that store information related to a memory, and whose reactivation is part of memory recall.

developing brain, as large-scale map plasticity is restricted to critical periods. Nevertheless, even in the adult brain, presentation of appropriate tetanization patterns can repeatedly reverse synaptic weights and de-potentiate synapses that previously underwent LTP, or de-depress synapses that previously underwent LTD (Han *et al.*, 2007). This feature shows that synaptic plasticity is a dynamic process. LTP and LTD have the capacity to permanently set synaptic weights, but only in the absence of further stimulation.[4] Any subsequent activity pattern has the capability of readjusting synaptic weights, once the conditions set by plasticity rules are met.

Synaptic potentiation is the activity- and experience-dependent component behind engram formation. The term 'engram' describes a group of neurons, whose physiological properties have been altered in learning, and whose reactivation is necessary for memory recall (Figure 3.2). Basic rules for the conditions under which stabilization/potentiation would occur were proposed in 1949 by the Canadian psychologist Donald Hebb. In his influential book *The Organization of Behavior*, Hebb (1949) stated:

*When an axon of cell A is near enough to excite a cell B and repeatedly or persistently takes part in firing it, some growth process*

---

[4]It is also conceivable that synaptic weights 'fade' over time. For example, LTP levels may decline, unless potentiation is renewed. This phenomenon is likely to exist, but has not been studied in detail.

*or metabolic change takes place in one or both cells such that A's efficiency, as one of the cells firing B, is increased.*

A more popular version of this rule reads, *'Neurons that fire together wire together'.* While Hebb's rule does not explicitly mention the weakening of synapses, this effect was suggested years later by Gunther Stent (1973) to include a negative change in input weight at those synapses, whose activity is not predictive of the firing of the target neuron.

## Calcium Influx as a Trigger Signal for Long-term Potentiation (and Depression)

How are synaptic weights actually changed, and what cellular signaling events initiate these changes? The remainder of this chapter will focus on these mechanistic questions. Before delineating the molecular machinery step by step, allow me to give away the core concepts. In short, the strength of synaptic transmission (the synaptic weight) can be altered both presynaptically, by modifying transmitter release in synaptic terminals, or postsynaptically, by modifying the responsiveness of the target neuron to the same amount of transmitter being released. Most excitatory synapses in the mammalian brain rely on the transmitter glutamate. At synaptic terminals, glutamate is released into the synaptic cleft when an action potential propagates down the axon, and invades and depolarizes the terminal. Specific receptors on the target neuron bind glutamate, resulting in the opening of ion channels.[5] The basic type of glutamate receptor is called an AMPA receptor (named after the synthetic agonist α-amino-3-hydroxy-5-methyl-4-isoxazole-proprionate). In synaptic plasticity, synaptic weight can be changed[6] by a) altering the biophysical properties of AMPA receptors, such that their binding affinity for glutamate is higher, or their ion conductance is larger, and b) altering their expression density in the postsynaptic membrane (Figure 3.3). Both processes are steered by enzymes. At glutamatergic synapses in the neocortex and hippocampus, kinase

---

[5] Such transmitter receptors are called 'ionotropic receptors'.
[6] I will focus here exclusively on postsynaptic plasticity mechanisms.

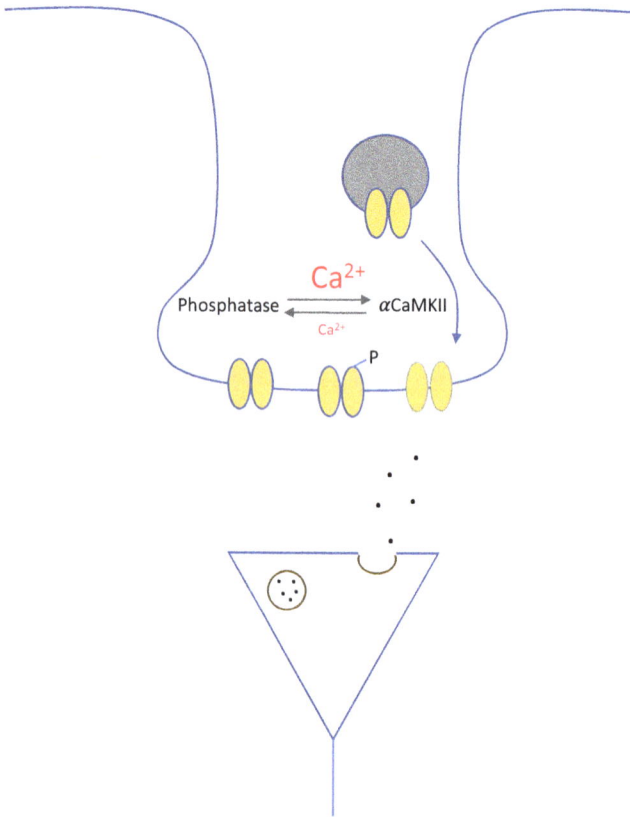

**Figure 3.3:** LTP depends on calcium signaling and kinase activation. An increase in the amplitude of postsynaptic responses to transmitter release (bottom) is achieved by the activation of kinases (here αCaMKII) that phosphorylate AMPA receptors (yellow) and enhance their membrane expression or alter their glutamate affinity or ion conductance. Phosphatase activation reverses these effects and causes LTD. Both processes are steered by activity-evoked calcium transients of different amplitude and duration.

activity phosphorylates specific amino acids within the AMPA receptor complex (adds phosphoryl groups) and alters its biophysical properties and membrane expression to enhance synaptic responses. In contrast, phosphatase activity dephosphorylates those amino acids and reverses these effects to lower synaptic responses. One of the most intriguing phenomena in cellular neurophysiology is that both potentiation and depression are driven by increases in the postsynaptic calcium concentration that may activate kinases or phosphatases (Lisman, 1989).

The experimental evidence for a role of calcium signals in the induction of both LTP and LTD is manifold:

1) Microscopic calcium imaging from pyramidal neurons using fluorescent calcium indicators that were injected into these cells show that both LTP and LTD inducing synaptic stimuli, tested in the same recordings, enhance the cytoplasmic calcium concentration (Hansel et al., 1997; Cormier et al., 2001; Nevian and Sakmann, 2006).
2) Loading neurons with substances that bind calcium and prevent it from activating enzymes[7] blocks LTP and LTD (e.g. Malenka et al., 1988, Hirsch and Crepel, 1990; Mulkey and Malenka, 1992; Bolshakov and Siegelbaum, 1994).
3) LTP and LTD can be triggered by photolysis of the photolabile calcium chelator nitr-5, a substance that binds calcium and releases it upon UV light exposure (Malenka et al., 1988; Neveu and Zucker, 1996).

These experiments show that the cell-internal calcium concentration $[Ca^{2+}]_i$ increases during stimuli that induce LTP or LTD (imaging), that elevated $[Ca^{2+}]_i$ is needed for induction (calcium chelators), and that experimentally enhancing $[Ca^{2+}]_i$ triggers LTP or LTD, even in the absence of synaptic activation (photo-release). The key question arising from these observations is this: How is it possible that the same trigger signal—calcium influx—can initiate both LTP and LTD, plasticity mechanisms that alter synaptic weight in *opposite* ways? I already pointed toward a possible answer that was suggested by the late John Lisman at Brandeis University (1989): calcium transients have different parameters, such as peak amplitude, duration, and localization. Differences in these parameters can distinguish one calcium signal from another and initiate different signaling pathways. In Lisman's model, a low-amplitude calcium signal activates phosphatases. This signaling pathway has a high affinity for calcium and thus gets activated at low calcium concentrations. In contrast, some

[7]These are the so-called calcium chelators, such as BAPTA [bis(2-aminophenoxy)ethane-N,N,N',N'-tetraacetate].

kinases such as $\alpha$CaMKII have a lower affinity for calcium and only become activated at higher calcium concentrations. However, once activated, this kinase pathway overpowers the phosphatase pathway, and the kinase-driven potentiation dominates (Figure 3.3). Formally, Lisman's proposal is based on the two-threshold synaptic modification model suggested by Bienenstock *et al.* for the developing visual cortex (BCM rule; 1982): a first threshold level of postsynaptic depolarization—which translates into calcium signal amplitudes—needs to be reached for synaptic depression. A higher, second threshold needs to be reached for potentiation (Figure 3.4). Experimental evidence (imaging/manipulation of calcium influx or cytosolic concentration) indeed supports the idea that bidirectional synaptic plasticity is governed by a calcium-based two-threshold mechanism (Mulkey and Malenka, 1992; Cummings *et al.*, 1996; Hansel *et al.*, 1997; Cormier *et al.*, 2001).

Synaptic plasticity and its underlying molecular machinery have initially been described in the hippocampus and neocortex. Subsequent studies, based on recordings from other brain areas, have shown that the induction rules follow similar schemes, but with potential differences. One of the best studied synapses outside of the neocortex and hippocampus is the synapse between Purkinje cells in the cerebellum and their so-called parallel fiber inputs, which arise from granule cells. At these synapses, LTD results from coactivation of parallel fibers with the second type of excitatory synapse onto Purkinje cells, the climbing fiber, which originates in the inferior olive. In contrast, PF activity alone causes LTP. As illustrated in Figure 3.4, there is a higher calcium threshold for LTD than for LTP, thus providing an example of a threshold mechanism that operates in an inverse mode, relative to its BCM counterpart (Coesmans *et al.*, 2004; Piochon *et al.*, 2016).

The threshold mechanisms discussed here are based on the amplitude of the calcium signal. What about other parameters of a calcium transient, such as location and duration? The need for calcium transients in specific locations certainly is important due to restrictions in the location of relevant calcium sensors such as CaMKII. These calcium sensors detect and use the amplitude and

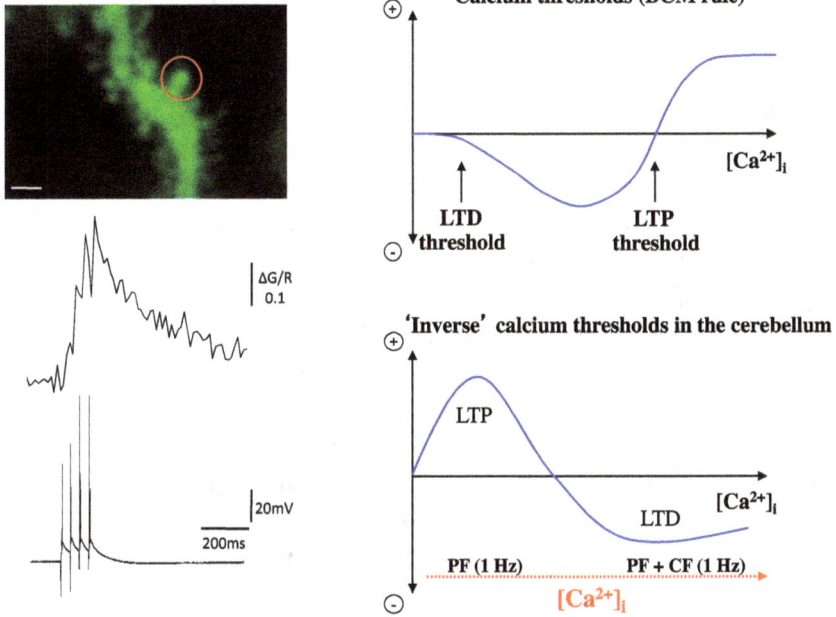

**Figure 3.4:** Two-threshold models for synaptic modification. The left side of the figure shows a measurement of a synaptically evoked calcium transient in a dendritic spine of a cerebellar Purkinje cell. Spines are dendritic protrusions that are the major contact sites for glutamatergic synapses. Scale bar: 1 μm. The traces below show the fluorescence signal (measuring the calcium transient in a dendritic spine) and the electrical response (measured in the cell body). The right side of the figure shows the classic, modified BCM rule. A higher $[Ca^{2+}]_i$ elevation needs to be reached for LTP relative to LTD induction, that is, the LTP threshold is higher than the LTD threshold (top). In contrast, at cerebellar parallel fiber (PF) to Purkinje cell synapses, the LTD threshold is higher than the LTP threshold. Coactivation of the climbing fiber (CF) input increases spine calcium signaling and helps to reach the threshold for induction of LTD.

duration of the signal for activation. Following this argument, the type of calcium source should be of secondary importance. Yet, calcium sources can create preferred patterns of calcium influx in calcium nanodomains. For example, the activation of N-methyl-D-aspartate (NMDA) receptors is required for hippocampal and neocortical plasticity under most experimental conditions (Bliss and Collingridge, 1993; Malenka and Nicoll, 1993). Like AMPA receptors, NMDA receptors bind glutamate. In contrast to AMPA receptors, though, NMDA receptors need postsynaptic depolarization in addition to glutamate binding in order to open, making them coincidence detectors of pre- and postsynaptic activity. Their associated ion channels are permeable to $Na^+$ and $Ca^{2+}$ ions, and thus they provide a major source of activity-dependent calcium influx in neurons.[8] The proximity of NMDA receptors and CaMKII in dendritic spines puts NMDA receptors into an optimal position to control nanodomain calcium signaling and CaMKII activation (Lee *et al.*, 2009). However, LTP can also be activated perfectly well when NMDA receptors are blocked, as long as other calcium sources, such as voltage-gated calcium channels, are active (Grover and Teyler, 1990). These results show that preferred sources of calcium influx may exist, but that ultimately the route of entry does not matter as long as the right calcium amplitude is reached at the location of the critical calcium sensor. Similarly, it has been shown that for LTP and LTD induction at glutamatergic synapses onto L2/3 pyramidal neurons in somatosensory cortex, two different types of calcium entry routes are preferred: (1) activation of NMDA receptors and voltage-dependent calcium channels for LTP, and (2) calcium release from internal calcium stores, a mechanism triggered by metabotropic glutamate receptors (mGluRs), for LTD. Blockade of mGluRs prevents LTD, but LTP can be induced if the calcium transient is sufficiently large (Nevian and Sakmann, 2006). Again, a scenario emerges in which under physiological conditions, specific calcium sources are recruited to trigger LTP or LTD at high probability. Nevertheless, from a mechanistic point of view, the amplitude thresholds appear to ultimately govern bidirectional plasticity.

The duration of calcium transients plays a role in synaptic plasticity as well. In the so-called 'leaky integrator model' of plasticity, the calcium threshold for cerebellar LTD becomes lower with a longer

---

[8]More recently, it has been suggested that NMDA receptors have a metabotropic function, that is, they do not influence plasticity via direct calcium influx, but by activating a biochemical signaling cascade in the target neuron (Nabavi *et al.*, 2013). How this result aligns with the classical view of NMDA channels described here, as a direct source of calcium influx, is the subject of current investigations.

presentation of the calcium elevation (Tanaka *et al.*, 2007; Chimal and De Schutter, 2018; Titley *et al.*, 2019). This finding shows that calcium-dependent processes are threshold-controlled, but that these thresholds do not assume absolute values. Instead, they can be modified depending on signal duration, that is, the amplitude thresholds slide depending on other parameters of the calcium signal (Figure 3.5). Similarly, my own research team at the University of Chicago has found that the calcium amplitude thresholds for cerebellar LTD and LTP slide depending on the frequency of parallel fiber stimulation. Both LTD and LTP can be induced by 1Hz as well as 100Hz activation protocols. At both frequencies, coactivation of the climbing fiber input promotes LTD (Piochon *et al.*, 2016). However, at 100Hz the calcium threshold for LTD induction is higher than the LTD threshold at 1Hz. Accordingly, 100Hz stimulation in the absence of climbing fiber coactivation causes LTP, although it elicits higher calcium transients than parallel fiber stimulation at 1Hz when it is paired with climbing fiber activity. Thus, the calcium amplitude obtained with this 100Hz stimulation does not reach the now elevated calcium

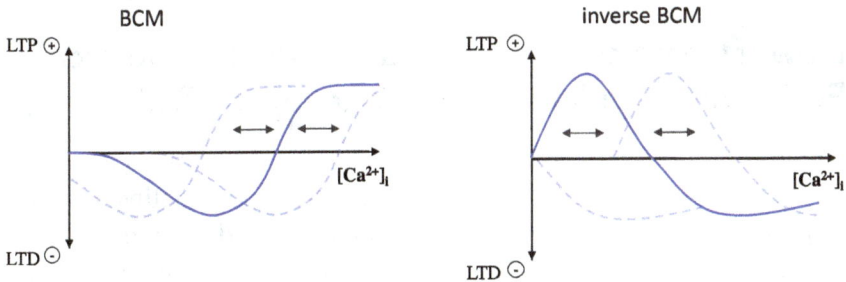

**Figure 3.5:** In both the BCM (left) and inverse BCM model (right) of threshold-controlled plasticity, the thresholds (or the point of sign reversal) slide depending on the activation history of the synapse (Bienenstock *et al.*, 1982; Cooper and Bear, 2012), or depending on parameters of the calcium signal experienced during synaptic tetanization itself (Piochon *et al.*, 2016). On the right, I depicted a scenario where the threshold shift to the left is sufficiently strong to make one direction of plasticity (here LTP) disappear. Such a shift might result when synaptic activation is set to zero and $[Ca^{2+}]_i$ is elevated by photolysis of calcium chelators. In that scenario, it is predicted that calcium elevation leads to LTD, but LTP is not observed. This prediction indeed aligns with the findings reported by Tanaka *et al.* (2007).

threshold for LTD. This phenomenon allows synapses to undergo potentiation or depression regardless of the synaptic activation frequency, but under control of the more crucial climbing fiber signal. The sliding threshold implies that the *absolute* calcium amplitude is not the critical factor in LTD and LTP control. However, the *relative* thresholds remain unchanged, that is, still at every state that the thresholds assume, more calcium is needed for LTD than for LTP. We calculated that the threshold shift is so extreme that around 0.4µM $[Ca^{2+}]_i$ is associated with LTD induction using 1Hz stimulation, while it is >7µM at 100Hz.

In these last paragraphs, I have focused on the role of calcium signaling in synaptic plasticity, because calcium transients in dendritic spines assume a unique role in determining whether synapses potentiate or depress. I have also outlined in detail why I believe that a two-threshold, amplitude-based mechanism is at the core of this decision process, even if several reports indicate that the source and location of calcium signals and their duration are additional factors. What happens downstream of calcium signaling? What are the calcium-driven processes that ultimately adjust synaptic response strength?

# Downstream of Calcium Transients: Biochemical Pathways Regulating AMPA Receptor Trafficking

The insertion of AMPA receptors into synaptic membranes and their removal ('internalization') are efficient means of potentiating synaptic responses or depressing response strength. Indeed, as illustrated in Figure 3.3, receptor trafficking is a core component of synaptic plasticity. Let us have an in-depth look at these processes at glutamatergic synapses in both the cerebellum and the hippocampus/neocortex.[9]

AMPA receptors are heteromeric complexes of the four subunits GluA1 to GluA4.[10] In hippocampal CA1 pyramidal cells, the majority of AMPA receptors consists of GluA1/GluA2 and GluA2/GluA3

---

[9]This paragraph is adapted from Jörntell and Hansel (2006).
[10]Heteromeric complexes are composed of different, rather than the same subunits (unlike homomeric complexes, which consist of multiples of the same subunit).

complexes. In GluA1/GluA2 complexes, GluA1 dominates the trafficking behavior (Song and Huganir, 2002). In contrast, Purkinje cells only weakly express GluA1 subunits (Baude *et al.*, 1994), and GluA2/GluA3 complexes constitute the majority of AMPA receptors. As we will see, this difference in AMPA receptor composition explains the difference in receptor trafficking and, ultimately, the inverse calcium threshold rules found at these different synapses (Figure 3.6).

LTD at parallel fiber synapses is mediated by the endocytosis (removal from the cell membrane) of GluA2 receptor subunits (Wang and Linden, 2000). This internalization requires PKC-dependent phosphorylation of GluA2 at amino acid serine Ser-880 (Chung *et al.*, 2003). The subsequent endocytotic processes involve complex interactions with proteins that are specialized for receptor shuffling. A key

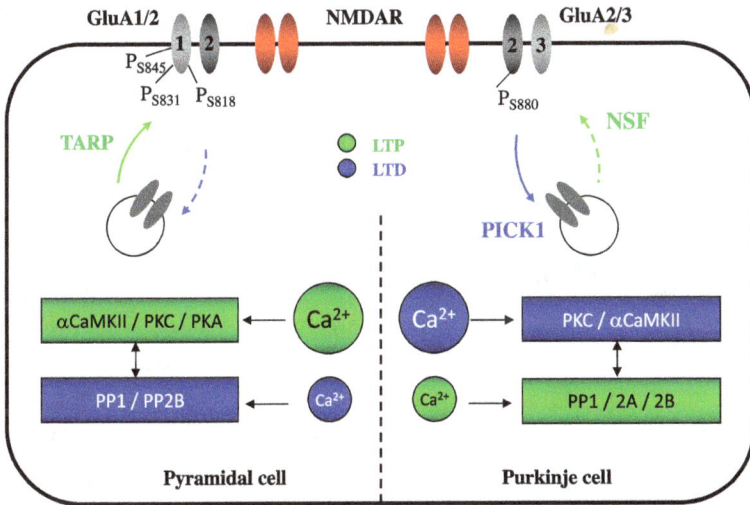

**Figure 3.6:** Comparison of LTP and LTD induction cascades at hippocampal (left) and cerebellar synapses (right). LTD induction pathways are shown in blue, and LTP induction pathways in green. For simplicity, no synaptic terminals are shown. One schematic postsynaptic compartment is used to illustrate LTP and LTD pathways in both pyramidal cells and Purkinje cells. The dashed line separating the 'pyramidal cell' from the 'Purkinje cell' does not continue all the way to the top, as PICK1-controlled GluA2 endocytosis (right) occurs in both types of neurons. Key: TARP: transmembrane AMPA receptor regulatory protein, e.g. stargazin. Other abbreviations are explained within the text. This figure is adapted from Jörntell and Hansel (2006).

step is the unbinding of GluA2 from the glutamate-receptor-interacting protein GRIP1 (Dong *et al.*, 1997) and subsequent binding to protein interacting with c-kinase 1 (PICK1; Xia *et al.*, 1999; Chung *et al.*, 2000), resulting in GluA2 internalization (Xia *et al.*, 2000). GRIP1 and PICK1 are synaptic proteins that bind to receptor subunits in a manner that can be controlled by kinase activity. This is precisely the role of PKC phosphorylation in this signaling pathway. It disrupts the binding of GluA2 to GRIP1, and frees the receptor subunit up for endocytosis. This sequence of events is typical for cellular signaling cascades. Phosphorylation alters protein conformation and the capacity to bind to, and interact with, specific protein partners. CaMKII plays a different role in this process. This kinase does not directly engage in the receptor internalization process, but instead inhibits activity of an opposing phosphatase pathway that might undo the consequences of PKC activation if left unchecked (Kawaguchi and Hirano, 2013). Purkinje cells express NMDA receptors at climbing fiber synaptic inputs (Piochon *et al.*, 2007; Renzi *et al.*, 2007). These NMDA receptors are able to contribute to the overall calcium signal and to LTD induction (Piochon *et al.*, 2010), likely through pathways that include the activation of CaMKII (Hansel *et al.*, 2006). LTP at parallel fiber synapses involves the addition (insertion) of GluA2 to the membrane, via GluA2 interaction with N-ethylmaleimide-sensitive factor (NSF; Kakegawa and Yuzaki, 2005). Many of the studies on AMPA receptor trafficking that are discussed here have been performed by Richard Huganir and David Linden at Johns Hopkins University. This work has subsequently enabled comparisons between molecular plasticity machineries at different types of synapses. I had the fortune to witness these developments as a postdoctoral fellow in David Linden's laboratory (during the years 1997 to 2000), an experience that shapes my thinking about synapse physiology to this day.

In contrast to these cerebellar synapses, GluA1 subunit trafficking plays a dominant role in the hippocampus. The phosphorylation events involved in CA1 hippocampal LTP induction are complex: phosphorylation of GluA1 at the PKA phosphorylation site Ser-845 seems to act as a 'priming' step for membrane insertion (Esteban *et al.*, 2003). This membrane integration of GluA1/GluA2 complexes is associated with CaMKII-mediated phosphorylation of Ser-831 on

GluA1 (Lee *et al.*, 2000) as well as the phosphorylation of additional support proteins, such as stargazin, a small transmembrane AMPA receptor regulatory protein (TARP; Nicoll *et al.*, 2006).

Ser-831 phosphorylation does not affect receptor trafficking itself (Hayashi *et al.*, 2000), but might increase receptor conductance (Derkach *et al.*, 1999). GluA1 membrane insertion and LTP are driven by a PKC-mediated phosphorylation at Ser-818 (Boehm *et al.*, 2006). A picture emerges, in which very specific aspects of receptor subunit trafficking are regulated by the phosphorylation of particular subunit phosphorylation sites.[11] The same holds true for subunit internalization that is required for LTD. At hippocampal synapses, GluA1 dephosphorylation events are associated with LTD (Lee *et al.*, 2000, 2003). Phosphatase inhibition blocks GluA1 internalization (Beattie *et al.*, 2000). These findings suggest that hippocampal LTP and LTD are dominated by GluA1 receptor subunits. Both LTP and LTD are indeed present in genetically engineered mice that are deficient of GluA2 receptors (Jia *et al.*, 1996). The role of GluA2 trafficking at hippocampal synapses therefore remains unclear. One popular hypothesis states that the membrane insertion of GluA2 is not driven by synaptic activity, but rests on NSF-dependent, continuous delivery, that is, it is an activity-independent process (Shi *et al.*, 2001). I have argued elsewhere (Hansel, 2005)[12] that it is unlikely that this process is strictly activity-independent. This is because the binding of NSF to GluA2, which precedes membrane insertion, is triggered by nitric oxide (NO) (Huang *et al.*, 2005; Kakegawa and Yuzaki, 2005), thus adding a signaling step that is activity-dependent in itself. NO is an atypical messenger molecule that is not released via synaptic vesicles and has no specific membrane receptors. It is produced by NO synthase through a calcium/calmodulin-dependent process (Garthwaite *et al.*, 1988; Bredt *et al.*, 1990). At some synapses, it acts as a retrograde messenger, that is, it is produced postsynaptically and

---

[11] A recent paper suggests that LTP may also result from membrane insertion of unmodified AMPA receptors, based on changes in the availability of 'receptor slots' (Diaz-Alonso *et al.*, 2020).

[12] In a commentary published in the *Proceedings of the National Academy of Sciences* (Vol 102, 2005).

diffuses to the presynaptic terminal. At others, including cerebellar synapses, it provides an anterograde (forward) signaling mechanism (Hansel *et al.*, 1992). At such synapses, its activity (calcium influx)-dependent production will impact NSF signaling and AMPA receptor trafficking. Moreover, the membrane removal of GluA2 at hippocampal synapses is activity-dependent and involved in LTD. GluA2 endocytosis is governed by the same PKC-dependent phosphorylation at Ser-880 at these hippocampal synapses (Man *et al.*, 2000; Seidenman *et al.*, 2003) as it is at cerebellar synapses (Wang and Linden, 2000; Chung *et al.*, 2003). These findings support a role of activity-dependent GluA2 trafficking in synaptic plasticity and learning.

Taken together, the picture of a mirror image-like arrangement of induction cascades between CA1 hippocampal and cerebellar plasticity emerges very clearly when comparing the calcium signaling requirements and the kinase/phosphatase involvement (Figure 3.6). The observed inverse induction cascades indicate that synaptic memories are governed by different mechanisms at different types of glutamatergic synapses. An interesting observation is that, at hippocampal and neocortical synapses, potentiation and depression are not regulated by a single 'toggle' switch mechanism (Figure 3.7). Rather, GluA1-dominated AMPA receptor complexes (GluA1/GluA2) can be switched on and off, and so can GluA2-dominated complexes (GluA2/GluA3). The same is true for critical phosphorylation sites on the GluA1 subunit itself. There is not just one phosphorylation site that is either phosphorylated ('on') or not ('off'). Instead, there are at

**Figure 3.7:** Toggle switch arrangements to regulate synaptic weights.

least two phosphorylation sites that are separately modulated. These dual toggle-switch mechanisms imply that synaptic weight is regulated via potentiation and de-potentiation as well as depression and de-depression. In contrast, at cerebellar synapses single toggle switches determine synaptic weights (GluA2, regulated by phosphorylation of Ser-880). It remains to be determined what the functional consequences of these arrangements are, but it seems plausible that a dual toggle switch would allow for a more fine-grained adjustment of synaptic input weights.

Why is a painstaking characterization of synaptic function as portrayed here needed? The answer is that these molecular details are informative. They tell us about the similarity of synaptic changes in brain maturation and in adult learning (Chapter IV). They also tell us about how small, punctual abnormalities in synaptic function can have disastrous consequences for brain function, as it occurs in some brain developmental disorders (Chapter V).

# References

Baude, A., Molnar, E., Latawiec, D., McIlhinney, R.A., and Somogyi, P. (1994). Synaptic and nonsynaptic localization of the GluR1 subunit of the AMPA-type excitatory amino acid receptor in the rat cerebellum. *J. Neurosci.* 14, 2830–2843.

Beattie, E.C., Caroll, R.C., Yu, X., Morishita, W., Yasuda, H., von Zastrow, M., and Malenka, R.C. (2000). Regulation of AMPA receptor endocytosis by a signaling mechanism shared with LTD. *Nat. Neurosci.* 3, 1291–1300.

Bienenstock, E.L., Cooper, L.N., and Munro, P.W. (1982). Theory for the development of neuron selectivity: orientation specificity and binocular interaction in visual cortex. *J. Neurosci.* 2, 32–48.

Bliss, T.V.P., and Collingridge, G.L. (1993). A synaptic model of memory: long-term potentiation in the hippocampus. *Nature* 361, 31–39.

Boehm, J., Kang, M.G., Johnson, R.C., Esteban, J., Huganir, R.L., and Malinow, R. (2006). Synaptic incorporation of AMPA receptors during LTP is controlled by a PKC phosphorylation site on GluR1. *Neuron* 51, 213–225.

Bolshakov, V.Y., and Siegelbaum, S.A. (1994). Postsynaptic induction and presynaptic expression of hippocampal long-term depression. *Science* 264, 1148–1152.

Bredt, D.S., Hwang, P.M., and Snyder, S.H. (1990). Localization of nitric oxide synthase indicating a neural role for nitric oxide. *Nature* 347, 768–770.

Chimal, C.G., and De Schutter, E. (2018). Ca2+ requirements for long-term depression are frequency sensitive in Purkinje cells. *Front. Mol. Neurosci.* 11, 438.

Chung, H.J., Steinberg, J.P., Huganir, R.L., and Linden, D.J. (2003). Requirement of AMPA receptor GluR2 phosphorylation for cerebellar long-term depression. *Science* 300, 1751–1755.

Chung, H.J., Xia, J., Scannevin, R.H., Zhang, X., and Huganir, R.L. (2000). Phosphorylation of the AMPA receptor subunit GluR2 differentially regulates its interaction with PDZ domain-containing proteins. *J. Neurosci.* 20, 7258–7267.

Coesmans, M., Weber, J.T., De Zeeuw, C.I., and Hansel, C. (2004). Bidirectional parallel fiber plasticity in the cerebellum under climbing fiber control. *Neuron* 44, 691–700.

Cooper, L.N., and Bear, M.F. (2012). The BCM theory of synapse modification at 30: interaction of theory with experiment. *Nat. Rev. Neurosci.* 13, 798–810.

Cormier, R.J., Greenwood, A.C., and Connor, J.A. (2001). Bidirectional synaptic plasticity correlated with the magnitude of dendritic calcium transients above a threshold. *J. Neurophysiol.* 85, 399–406.

Cummings, J.A., Mulkey, R.M., Nicoll, R.A., and Malenka, R.C. (1996). Ca2+ signaling requirements for long-term depression in the hippocampus. *Neuron* 16, 825–833.

Derkach, V., Barria, A., and Soderling, T.R. (1999). Ca2+/calmodulin-kinase II enhances channel conductance of alpha-amino-3-hydroxy-5-methyl-4-isoxazoleproprionate type glutamate receptors. *Proc. Natl. Acad. Sci. USA* 96, 3269–3274.

Diaz-Alonso, J., Morishita, W., Incontro, S., Simms, J., Holtzmann, J., Gill, M., Mucke, L., Malenka, R.C., and Nicoll, R.A. (2020). Long-term potentiation is independent of the C-tail of the GluA1 AMPA receptor subunit. *eLife* 9, e58042.

Dong, H., O'Brien, R.J., Fung, E.T., Lanahan, A.A., Worley, P.F., and Huganir, R.L. (1997). GRIP: a synaptic PDZ domain-containing protein that interacts with AMPA receptors. *Nature* 386, 279–284.

Esteban, J.A., Shi, S.H., Wilson, C., Nuriya, M., Huganir, R.L., and Malinow, R. (2003). PKA phosphorylation of AMPA receptor subunits controls synaptic trafficking underlying plasticity. *Nat. Neurosci.* 6, 136–143.

Feldman, D.E., and Brecht, M. (2005). Map plasticity in somatosensory cortex. *Science* 310, 810–815.

Garthwaite, J., Charles, S., and Chess-Williams, R. (1988). Endothelium-derived relaxing factor release on activation of NMDA receptors suggests role as intercellular messenger in the brain. *Nature* 336, 385–387.

Grover, L.M., and Teyler, T.J. (1990). Two components of long-term potentiation induced by different patterns of afferent activation. *Nature* 347, 477–479.

Han, V.Z., Zhang, Y., Bell, C.C., and Hansel, C. (2007). Synaptic plasticity and calcium signaling in Purkinje cells of the central cerebellar lobes of mormyrid fish. *J. Neurosci.* 27, 13499–13512.

Hansel, C. (2005). When the B-team runs plasticity: GluR2 receptor trafficking in cerebellar long-term potentiation. *Proc. Natl. Acad. Sci. USA* 102, 18245–18246.

Hansel, C., Artola, A., and Singer, W. (1997). Relation between dendritic Ca2+ levels and the polarity of synaptic long-term modifications in rat visual cortex neurons. *Eur. J. Neurosci.* 9, 2309–2322.

Hansel, C., Batchelor, A., Cuenod, M., Garthwaite, J., Knöpfel, T., and Do, K.Q. (1992). Delayed increase of extracellular arginine, the nitric oxide precursor, following electrical white matter stimulation in rat cerebellar slices. *Neurosci. Lett.* 142, 211–214.

Hansel, C., de Jeu, M., Belmeguenai, A., Houtman, S.H., Buitendijk, G.H., Andreev, D., De Zeeuw, C.I., and Elgersma, Y. (2006). αCaMKII is essential for cerebellar LTD and motor learning. *Neuron* 51, 835–843.

Hayashi, Y., Shi, S.H., Esteban, J.A., Piccini, A., Poncer, J.C., and Malinow, R. (2000). Driving AMPA receptors into synapses by LTP and CaMKII: requirement for GluR1 and PDZ domain interaction. *Science* 287, 2262–2267.

Hebb, D.O. (1949). *The Organization of Behavior* (New York: Wiley).

Hirsch, J.C., and Crepel, F. (1992). Postsynaptic calcium is necessary for the induction of LTP and LTD of monosynaptic EPSPs in prefrontal neurons: an in vitro study in the rat. *Synapse* 10, 173–175.

Huang, Y., Man, H.Y., Sekine-Aizawa, Y., Han, Y., Juluri, K., Luo, H., Cheah, J., Loewenstein, C., Huganir, R.L., and Snyder, S.H. (2005). S-nitrosylation of N-ethylmaleimide sensitive factor mediates surface expression of AMPA receptors. *Neuron* 46, 533–540.

Hunt, D.L., Yamoah, E.N., and Krubitzer, L. (2006). Multisensory plasticity in congenitally deaf mice: how are cortical areas functionally specified? *Neuroscience* 139, 1507–1524.

Jia, Z., Agopyan, N., Miu, P., Xiong, Z., Henderson, J., Gerlai, R., Taverna, F.A., Velumian, A., MacDonald, J., Carlen, P., Abramow-Newerly, W., and Roder, J. (1996). Enhanced LTP in mice deficient in the AMPA receptor GluR2. *Neuron* 17, 945–956.

Jörntell, H., and Hansel, C. (2006). Synaptic memories upside down: bidirectional plasticity at cerebellar parallel fiber—Purkinje cell synapses. *Neuron* 52, 227–238.

Kahn, D.M., and Krubitzer, L. (2002). Massive cross-modal cortical plasticity and the emergence of a new cortical area in developmentally blind mammals. *Proc. Natl. Acad. Sci. USA* 99, 11429–11434.

Kakegawa, W., and Yuzaki, M. (2005). Novel mechanism underlying AMPA receptor trafficking during cerebellar long-term potentiation. *Proc. Natl. Acad. Sci. USA* 102, 17846–17851.

Kawaguchi, S.Y., and Hirano, T. (2013). Gating of long-term depression by Ca2+/calmodulin-dependent protein kinase II through enhanced cGMP signaling in cerebellar Purkinje cells. *J. Physiol.* 591, 1707–1730.

Lee, H.K., Barbarosie, M., Kameyama, K., Bear, M.F., and Huganir, R.L. (2000). Regulation of distinct AMPA receptor phosphorylation sites during bidirectional synaptic plasticity. *Nature* 405, 955–959.

Lee, H.K., Takamiya, K., Han, J.S., Man, H., Kim, C.H., Rumbaugh, G., Yu, S., Ding, L., He, C., Petralia, R.S., Wenthold, R.J., Gallagher, M., and Huganir, R.L. (2003). Phosphorylation of the AMPA receptor GluR1 subunit is required for synaptic plasticity and retention of spatial memory. *Cell* 112, 631–642.

Lee, S.J., Escobedo-Lozoya, Y., Szatmari, E.M., and Yasuda, R. (2009). Activation of CaMKII in single dendritic spines during long-term potentiation. *Nature* 458, 299–304.

Lisman, J. (1989). A mechanism for the Hebb and the anti-Hebb processes underlying learning and memory. *Proc. Natl. Acad. Sci. USA* 86, 9574–9578.

Malenka, R.C., Kauer, J.A., Zucker, R.S., and Nicoll, R.A. (1988). Postsynaptic calcium is sufficient for potentiation of hippocampal synaptic transmission. *Science* 242, 81–84.

Malenka, R.C., and Nicoll, R.C. (1993). NMDA-receptor-dependent synaptic plasticity: multiple forms and mechanisms. *Trends Neurosci.* 16, 521–527.

Man, Y.H., Lin, Y.W., Ju, W.H., Ahamadian, G., Liu, L., Becker, L.E., Sheng, M., and Wang, Y.T. (2000). Regulation of AMPA receptor-mediated synaptic transmission by clathrin-dependent receptor internalization. *Neuron* 25, 649–662.

Mulkey, R.M., and Malenka, R.C. (1992). Mechanisms underlying induction of homosynaptic long-term depression in area CA1 of the hippocampus. *Neuron* 9, 967–975.

Nabavi, S., Kessels, H.W., Alfonso, S., Aow, J., Fox, R., and Malinow, R. (2013). Metabotropic NMDA receptor function is required for NMDA receptor-dependent long-term depression. *Proc. Natl. Acad. Sci. USA* 110, 4027–4032.

Neveu, D., and Zucker, R.S. (1996). Postsynaptic levels of [Ca2+]i needed to trigger LTD and LTP. *Neuron* 16, 619–629.

Nevian, T., and Sakmann, B. (2006). Spine Ca2+ signaling in spike-timing-dependent plasticity. *J. Neurosci.* 26, 11001–11013.

Nicoll, R.A., Tomita, S., and Bredt, D.S. (2006). Auxilliary subunits assist AMPA-type glutamate receptors. *Science* 311, 1253–1256.

Piochon, C., Irinopoulou, T., Brusciano, D., Bailly, Y., Mariani, J., and Levenes, C. (2007). NMDA receptor contribution to the climbing fiber response in the adult mouse Purkinje cell. *J. Neurosci.* 27, 10797–10809.

Piochon, C., Levenes, C., Ohtsuki, G., and Hansel, C. (2010). Purkinje cell NMDA receptors assume a key role in synaptic gain control in the mature cerebellum. *J. Neurosci.* 30, 15330–15335.

Piochon, C., Titley, H.K., Simmons, D.H., Grasselli, G., Elgersma, Y., and Hansel, C. (2016). Calcium threshold shift enables frequency-independent control of plasticity by an instructive signal. *Proc. Natl. Acad. Sci. USA* 113, 13221–13226.

Renzi, M., Farrant, M., and Cull-Candy, S.G. (2007). Climbing fibre-activation of NMDA receptors in Purkinje cells of adult mice. *J. Physiol.* 585, 91–101.

Seidenman, K.J., Steinberg, J.P., Huganir, R.L., and Malinow, R. (2003). Glutamate receptor subunit 2 serine 880 phosphorylation modulates synaptic transmission and mediates plasticity in CA1 pyramidal cells. *J. Neurosci.* 23, 9220–9228.

Semon, R. (1904). *Die Mneme als erhaltendes Prinzip im Wechsel des organischen Geschehens* (Leipzig: Wilhelm Engelmann).

Shi, S.H., Hayashi, Y., Esteban, J.A., and Malinow, R. (2001). Subunit-specific rules governing AMPA receptor trafficking to synapses in hippocampal pyramidal neurons. *Cell* 105, 331–343.

Song, I., and Huganir, R.L. (2002). Regulation of AMPA receptors during synaptic plasticity. *Trends Neurosci.* 25, 578–588.

Stent, G.S. (1973). A physiological mechanism for Hebb's postulate of learning. *Proc. Natl. Acad. Sci. USA* 70, 997–1001.

Tanaka, K., Khiroug, L., Santamaria, F., Doi, T., Ogasawara, H., Ellis-Davies, G.C., Kawato, M., and Augustine, G.J. (2007). Ca2+ requirements for cerebellar long-term synaptic depression: role for a postsynaptic leaky integrator. *Neuron* 54, 787–800.

Titley, H.K., Kislin, M., Simmons, D.H., Wang, S.S.H., and Hansel, C. (2019). Complex spike clusters and false-positive rejection in a cerebellar supervised learning role. *J. Physiol.* 597.16, 4387–4406.

Wang, Y.T., and Linden, D.J. (2000). Expression of cerebellar long-term depression requires postsynaptic clathrin-mediated endocytosis. *Neuron* 25, 635–647.

Xia, J., Chung, H.J., Wihler, C., Huganir, R.L., and Linden, D.J. (2000). Cerebellar long-term depression requires PKC-regulated interactions between GluR2/3 and PDZ domain containing proteins. *Neuron* 28, 499–510.

Xia, J., Zhang, X., Staudinger, J., and Huganir, R.L. (1999). Clustering of AMPA receptors by the synaptic PDZ domain-containing protein PICK1. *Neuron* 22, 179–187.

# 4 Across Time: Shared Synaptic Plasticity Machinery in the Developing and the Adult Brain

Memory makes the brain. This is the title that I selected for this book, because it emphasizes that our brains receive many signals that leave a lasting impact—they change the brain. This might be a somewhat crude analogy, but nevertheless it is true: in this way, the brain resembles a memory mattress, where every contact that is sufficiently strong leaves a mark in the form of a localized indentation that is only removed or altered when a new impact occurs. This 'change upon impact' concept is at the core of circuit learning. However, there are important distinctions in circuit learning at various stages of postnatal development. At all stages, synaptic depression and potentiation processes take place. Bidirectional synaptic plasticity is always available. And yet, pruning and synaptic depression constitute a dominating motif during childhood development, whereas synaptic potentiation leaves its mark on adult learning. Despite these pronounced differences in circuit learning at different stages of life, the molecular pathways involved in developmental pruning and adult LTD, respectively, and equally those involved in developmental synapse stabilization and adult LTP, are remarkably similar. This is an important point to make, and this chapter will focus on this longitudinal aspect of brain plasticity. It was Mark Bear at Brown University (he later moved his laboratory to the MIT), who first pointed out similarities between plasticity pathways in the developing and the adult brain (Bear and Rittenhouse, 1999). Why

**Figure 4.1:**   Molecular continuity in plasticity pathways. The color itself indicates the similarity of molecular pathways, whereas the color depth indicates the magnitude of plasticity. Typical postnatal time periods are shown for the mouse, where functional synapses on neuronal dendrites are present 7 days after birth (P7; varies with the type of synapse). At many synapses, developmental plasticity windows close at ~P21, and maturation is reached at >P60.

is this aspect of *molecular continuity* (Figure 4.1) so essential? Before I offer some answers to this question, let me point out that there are changes in the relative expression levels of signaling factors over time (Lohmann and Kessels, 2014). Thus, molecular continuity is a motif that is worth acknowledgement, but the continuity is certainly not absolute. We will make note of specific exceptions later in this chapter. However, there is continuity in many aspects of plasticity over time, and I will highlight these to work out important consequences of this phenomenon.[1] In the following, I will develop the argumentation in support of molecular continuity by discussing plasticity at three different synapses: cortical synapses in the visual pathways, parallel fiber (PF) and climbing fiber (CF) synapses onto cerebellar Purkinje cells, and the neuromuscular junction (NMJ).

## Pruning in the Visual Cortex Following Molecular Deprivation

The work by Hubel and Wiesel on connectivity changes during the critical period in the primary visual cortex of kittens (Wiesel and Hubel, 1963) has laid the groundwork for subsequent studies on synaptic pruning and its relationship to LTD. For this reason, I will begin with a discussion of activity-dependent pruning events in the primary visual cortex. As discussed in Chapter II, the majority of neurons clustered in a cortical column in the adult visual cortex respond preferentially to the stimulation of one eye, a phenomenon known as ocular dominance

---

[1] This chapter is in parts adapted from Piochon *et al.* (2016).

(Figure 2.2). The adult neuronal connectivity pattern results from developmental synaptic rewiring during the critical period, including the elimination of synapses serving the nondominant eye. The degree of plasticity during the critical period is remarkably high, as demonstrated by Hubel and Wiesel in their classic monocular deprivation studies: if one eye is deprived of visual input during the critical period, neurons lose responsiveness to input from this eye, and the ocular dominance segregation is heavily weighted toward input from the intact eye (Wiesel and Hubel, 1963). Synaptic depression initiated by monocular deprivation is an active process that requires residual activity in the visually deprived retina (Rittenhouse *et al.*, 1999). Thus, synaptic depression during developmental circuit plasticity shares activity dependence with LTD in the mature cortex. It was first suggested by Bienenstock, Cooper and Munro (in the same paper, in which they introduced the BCM threshold rule; Bienenstock *et al.*, 1982) that synaptic depression at inputs from the deprived eye initiates the loss of responsiveness to visual stimulation. This hypothesis has been experimentally supported. First, monocular deprivation by lid suture, which preserves residual activity in the retina, leads to a more pronounced depression of responses to input from the deprived eye than monocular inhibition by intraocular injection of tetrodotoxin (a $Na^+$ channel blocker that prevents action potential firing) (Rittenhouse *et al.*, 1999). Second, synaptic depression triggered by monocular deprivation occludes subsequent LTD induction (Heynen *et al.*, 2003), suggesting that monocular deprivation recruits LTD, and that LTD marks an early step in synaptic pruning.

The molecular pathways involved in LTD in the visual cortex and in deprivation-induced synaptic pruning show a strong overlap. NMDA receptor activation is required for both LTD (Kirkwood *et al.*, 1993; Kirkwood and Bear, 1994) and the ocular dominance shift following monocular deprivation (Kleinschmidt *et al.*, 1987; Bear *et al.*, 1990). Similarly, both LTD and ocular dominance plasticity are impaired by chronic blockade of type 5 metabotropic glutamate receptors (mGluR5; Sidorov *et al.*, 2015). In addition, both molecular deprivation and LTD cause the same changes in AMPA receptor phosphorylation and membrane expression, namely decreased phosphorylation of GluA1 at Ser-845, increased phosphorylation of GluA2 at Ser-880, and reduced surface expression of GluA1 and GluA2 subunits (Heynen *et al.*, 2003).

Thus, the ocular dominance shift following molecular deprivation and LTD share not only the specific types of transmitter receptors that need to be activated for the induction of the process but, even more convincingly, they also share molecular signatures of synapses with reduced responsiveness. Therefore, the overlap of pruning and LTD machineries can clearly be demonstrated at cortical synapses (Singer, 1995; Bear and Rittenhouse, 1999).

## The Molecular Machinery of Synaptic Depression at a Cerebellar Synapse

At glutamatergic synapses onto cerebellar Purkinje cells, the molecular pathways underlying both LTD at parallel fiber synapses (PF-LTD), and developmental synaptic pruning at CF synapses have been described at a level of detail that is not available from other synapses in the central nervous system (CNS). In fact, CF synapses are one of the few types of synapses in the CNS where developmental synaptic pruning can be readily assessed using electrophysiological techniques and where the number of inputs can be quantified by the number of discrete all-or-none steps in synaptic responses. At most other CNS synapses, it is more difficult to isolate responses evoked by single presynaptic neurons, and thus this quantification method cannot easily be applied. For these reasons, the study of cerebellar synapses has been a particularly fruitful approach to compare molecular machineries in developmental and adult plasticity.

LTD was first described as a potential cellular correlate of learning in the cerebellum. Here, PF-LTD onto Purkinje cells is thought to reduce the level of inhibition that the GABAergic Purkinje cells impose on their target cells in the cerebellar and vestibular nuclei, thus promoting activity in downstream motor control pathways.[2] LTD is triggered at those PF synapses that are co-active with the CF input (Ito *et al.*, 2014). PF-LTD depends on the activation of type 1 mGluRs (Figure 4.2; Aiba *et al.*, 1994; Conquet *et al.*, 1994; Ichise *et al.*,

---

[2]The cerebellum is primarily seen as a brain area involved in motor adjustment and learning, but may regulate cognitive processes as well through its interactions with cortical nonmotor areas.

2000). Group I mGluRs (consisting of mGluR 1 and 5; only mGluR1 is expressed at appreciable levels in Purkinje cells) couple to Gαq proteins that activate a signaling cascade, including phospholipase Cβ4 (PLCβ4) and protein kinase C (PKC; Knöpfel and Grandes, 2002). The activation of mGluR1 receptors specifically promotes LTD, but is not required for LTP induction (Belmeguenai et al., 2008). Consistent with an involvement of an mGluR1 signaling cascade, LTD is absent from Gαq knockout mice (Gnaq$^{-/-}$; Hartmann et al., 2004) and from PLCβ4 knockout mice (Plcb4$^{-/-}$; Hashimoto et al., 2001). PLC produces diacylglycerol and inositol-1,4,5-trisphosphate (IP$_3$) from phosphatidylinositol-4,5-bisphosphate, triggering the subsequent release of calcium from IP$_3$-sensitive intracellular stores. Both consequences of PLC activity—production of diacylglycerol and calcium release from IP$_3$-sensitive stores—promote the activation of PKC (Knöpfel and Grandes, 2002).

LTD is indeed prevented by PKC inhibitors (Linden and Connor, 1991; Chen et al., 1995) and is absent from transgenic mice that express the pseudosubstrate PKC inhibitor PKC$_{19-31}$ selectively in Purkinje cells (De Zeeuw et al., 1998). Subsequent studies showed that the phosphorylation of GluA2 subunits at Ser-880 by PKCα is a critical step in GluA2 endocytosis and LTD (Chung et al., 2003; Leitges et al., 2004). As Purkinje cells only weakly express GluA1 subunits, GluA2 at Ser-880 is the main PKC target in cerebellar LTD. In summary, PF-LTD requires the activation of an mGluR1- PLCβ4- PKCα signaling cascade that provides an early step in GluA2 endocytosis, which removes functional AMPA receptors from the postsynaptic density and mediates LTD.

# Climbing Fiber Pruning in the Developing Cerebellum

One of the best-studied examples of developmental synaptic pruning is the elimination of surplus CF inputs in the cerebellum. Purkinje cells are contacted at birth by usually three to five CFs, which are eliminated until, by the end of the third postnatal week, only one CF input remains in most Purkinje cells (Crepel et al., 1976). CF pruning is an

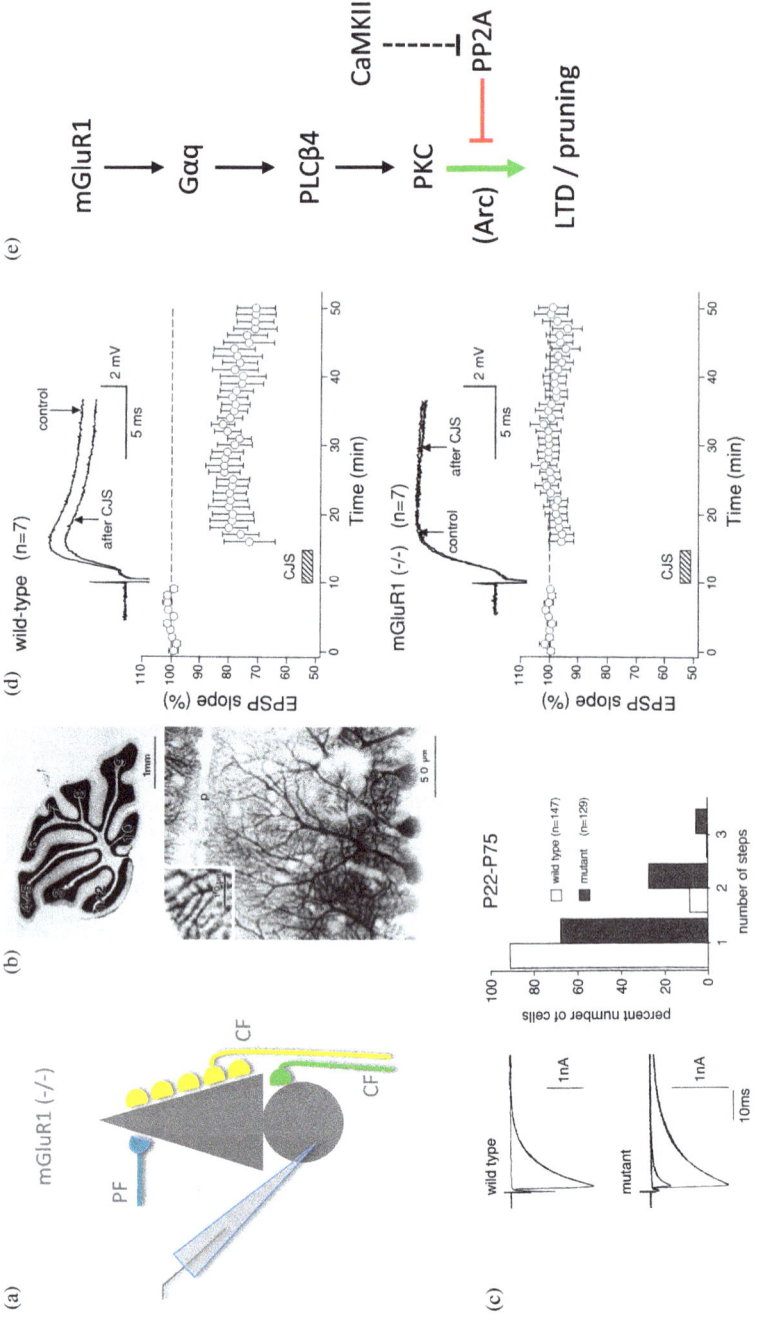

**Figure 4.2:** Deficient LTD and impaired CF synapse elimination in mGluR1 knockout mice. (a) Schematic of the patch-clamp configuration used to record from an mGluR1 knockout Purkinje cell that is illustrated with multiple CF innervation. (b) Gross anatomy of the cerebellum (top, Nissl staining) and morphology of Purkinje cell dendrites (bottom, calbindin immunostaining) are normal in mGluR1 knockout mice. (c) Persistent multiple CF innervation in adult mGluR1 knockout mice (P22-75). CF-mediated excitatory postsynaptic currents (EPSCs; left) and frequency distribution histogram showing the number of discrete CF EPSC steps at increasing stimulus strength (right), representing the number of CF inputs. (d) PF-LTD is deficient in mGluR1 knockout mice. In wild-type mice, excitatory postsynaptic potentials (EPSPs) elicited by PF stimulation undergo LTD after conjunctive PF and CF stimulation (CJS) at 1 Hz for 5 min (top). In contrast, PF-EPSPs are not depressed by CJS in mGluR1 knockout mice (bottom). All values are shown as mean ± SEM (standard error of the means). (e) mGluR1 signaling pathway. CaMKII activation contributes to LTD through an indirect blockade of protein phosphatase 2A (PP2A). CaMKII may similarly contribute to CF synaptic pruning, but this has not yet been verified. Arc: activity-regulated cytoskeleton-associated protein. Panels b and c adapted from Kano *et al.* (1997), d from Ichise *et al.* (2000).

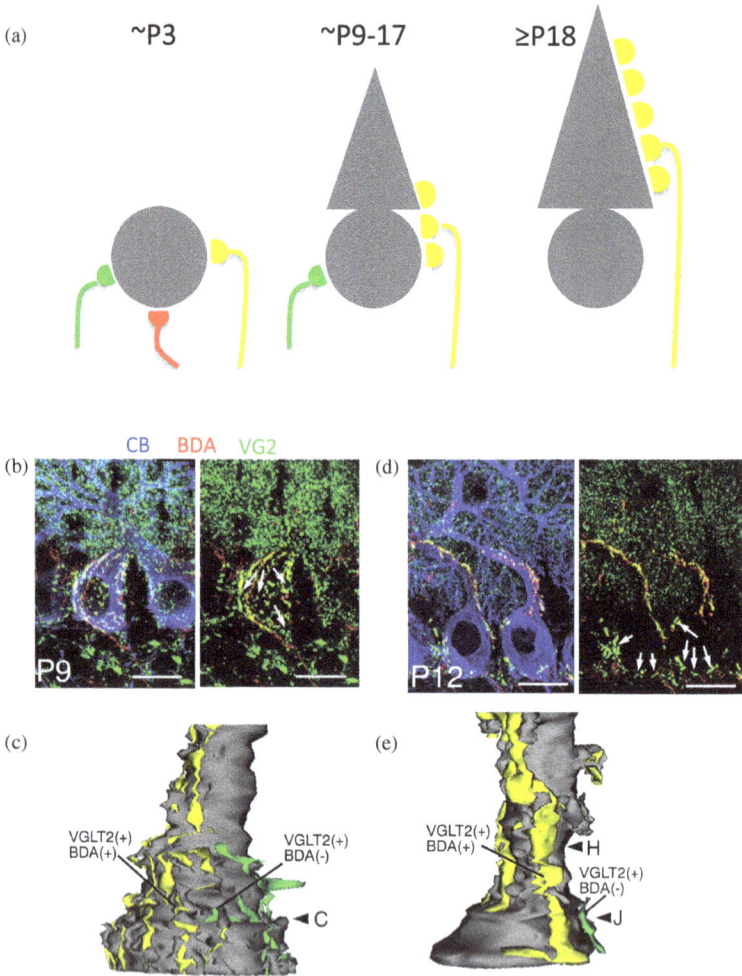

**Figure 4.3:** Developmental CF synapse elimination in the rodent cerebellum. (a) Schemes representing CF innervation of Purkinje cells at three stages of postnatal development. At around P3, the Purkinje cell soma is innervated by multiple CFs with similar synaptic strengths. At around P9–P17, a single winner CF extends its innervation territory from the soma to the growing neuronal dendrite, whereas the loser CFs maintain synapses on the soma. After P18, most of the somatic CF synapses are eliminated and a single winner innervates the dendrite, forming spines located on the primary dendrite. (b, d) Triple fluorescence labeling at P9 (b) and P12 (d) for biotinylated dextran amine (BDA, a tracer labeling a subset of CFs), VGluT2 (type 2 vesicular glutamate transporter, a marker of all CF terminals), and CB (calbindin, a Purkinje cell marker). (b) At P9, the soma is innervated by BDA and VGluT2 double-positive CFs (yellow puncta) and BDA-negative and VGluT2-positive CF terminals

active process and consists of distinct phases (Hashimoto and Kano, 2013). During the first postnatal week, differences in synaptic input strength of several CFs innervating the soma become noticeable, with a single CF input emerging that can be identified as the future 'winner' CF input on the basis of response strength and morphology; this is the functional differentiation phase (Hashimoto and Kano, 2003). It seems that the one CF input is selectively strengthened whose activity is most closely related to the occurrence of calcium spikes and calcium influx through P/Q-type voltage-gated calcium channels in the Purkinje cell dendrite (Hashimoto *et al.*, 2011; Kawamura *et al.*, 2013). Subsequently, the winner CF input begins to translocate to the dendrite while maintaining synapses on the soma (Figure 4.3; Hashimoto, Ichikawa, *et al.*, 2009; Carillo *et al.*, 2013).

The weaker CF inputs do not translocate to the dendrite and will be withdrawn in two distinct phases of CF elimination, an early and a late phase that occur largely in parallel with the translocation of the winner CF to the dendrite (Hashimoto, Ichikawa, *et al.*, 2009). This form of developmental plasticity is competitive, as reversal of the fate of a loser CF input has been observed following the ablation of the emerging winner CF input (Carillo *et al.*, 2013). The early phase of CF synapse elimination starts at around P7, just after the functional differentiation phase, and is independent of proper formation of PF to Purkinje cell synapses (Hashimoto, Yoshida, *et al.*, 2009). The late phase of CF synapse elimination starts around P12 and continues until around P17. This phase marks the ongoing synaptic competition process at CF synapses, and it critically depends on the proper formation and activity of PF synapses (Hashimoto, Yoshida, *et al.*, 2009). Eventually, the elimination of loser CF synapses is completed by

---

**Figure 4.3:** (*Facing page Continued*)   (arrows, green puncta), indicating innervation by two CFs. (d) At P12, the dendrites are innervated by BDA and VGluT2 double-positive CFs (yellow puncta) and the somata are contacted by BDA-negative and VGluT2-positive CFs (arrows, green puncta), indicating single strong CF inputs on Purkinje cell dendrites and additional weak CF inputs on the somata. (c, e) Three-dimensionally reconstructed image of CF innervation from serial electron microscope analysis of a Purkinje cell at P9 (c) and P12 (e). Scale bars in b and d, 10μm. Panels b–e are adapted from Hashimoto, Ichikawa, *et al.* (2009).

retraction. This process involves engulfment by glial cells for final digestion (Hashimoto and Kano, 2013). Observations at cortical synapses show that microglia are indeed involved in synaptic pruning during development and contribute to the clearance of cellular material even in the uninjured brain (Paolicelli *et al.*, 2011). Deficits in microglia recruitment or in the tagging of synapses for removal by microglia result in impaired synaptic pruning, enhanced seizure rates, and autism-resembling social behavior abnormalities (Chu *et al.*, 2010; Hoshiko *et al.*, 2012; Zhan *et al.*, 2014).

The molecular events involved in CF pruning, specifically the late phase of CF elimination, have been characterized in detail in a series of studies using genetically modified mice. In mutant mice lacking mGluR1 receptors (*Grm1$^{-/-}$* mice), the regression of surplus CFs slows at the end of the second postnatal week and remains incomplete even after P22, an age at which the elimination process is typically finished (Figure 4.2; Kano *et al.*, 1997; Levenes *et al.*, 1997). Purkinje cell-specific expression of an alpha isoform of mGluR1 (mGluR1$\alpha$) transgene rescues proper CF elimination (Ichise *et al.*, 2000). These data show that mGluR1 receptors are crucial to CF synapse elimination. Similarly, impairment of CF elimination at the end of the second postnatal week has been observed in mutant mice lacking the G protein G$\alpha$q (Gnaq$^{-/-}$ mice; Offermanns *et al.*, 1997), in mice lacking PLC$\beta$4 (*Plcb4$^{-/-}$*; Kano *et al.*, 1998; Hashimoto *et al.*, 2001), and in mice lacking PKC$\gamma$ (*Prkcg$^{-/-}$*; Kano *et al.*, 1995). Moreover, CF elimination is impaired in mice expressing the pseudosubstrate PKC inhibitor PKC$_{19-31}$ (De Zeeuw *et al.*, 1998). These findings show that the developmental elimination of surplus CF inputs depends, just as does PF-LTD, on an mGluR1-G$\alpha$q-PLC$\beta$4-PKC signaling cascade (Table 1.1).[3]

Signaling factors that are also shared by LTD and pruning, but are not a direct part of the mGluR1-G$\alpha$q-PLC$\beta$4-PKC signaling cascade, are the activation of $\alpha$CaMKII (Hansel *et al.*, 2006) and Arc, an immediate early gene product that promotes the endocytosis of

---

[3]The overlap in molecular machinery is not complete. LTD requires activation of PKC$\alpha$, whereas climbing fiber pruning requires the activation of PKC$\gamma$. In addition, LTD in the adult cerebellum depends on glutamate binding to NMDA receptors (Piochon *et al.*, 2010), which are not expressed before adulthood (Piochon *et al.*, 2007; Renzi *et al.*, 2007).

**Table 1.1:** Both LTD at PF synapses and the developmental elimination of surplus CF synapses are impaired (X) in genetically modified mice that, in most cases, were originally studied to examine the relationship between LTD and motor learning. The left column shows the targeted molecules.

| | LTD (PF) | Synaptic pruning (CF) | Reference |
|---|---|---|---|
| mGluR1 | X | X | Aiba A et al., *Cell* 79 (1994)<br>Conquet F et al., *Nature* 372 (1994)<br>Kano M et al., *Neuron* 18 (1997) |
| Gαq | X | X | Offermans S et al., *PNAS* 94 (1997)<br>Hartmann J et al., *J. Neurosci.* 24 (2004) |
| PLCβ4 | X | X | Kano M et al., *PNAS* 95 (1998)<br>Hashimoto K et al., *Mol. Neurobiol.* 23 (2001) |
| PKC | X | X | Kano M et al., *Cell* 83 (1995)<br>De Zeeuw C et al., *Neuron* 20 (1998)<br>Leitges M et al., *Neuron* 44 (2004) |
| αCaMKII | X | X | Hansel C et al., *Neuron* 51 (2006) |
| Arc | X | X | Smith-Hicks C et al., *Nat. Neurosci.* 13 (2010)<br>Mikuni T et al., *Neuron* 78 (2013) |

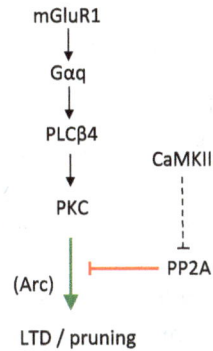

mGluR1 → Gαq → PLCβ4 → PKC → (Arc) → LTD / pruning; CaMKII ⊣ PP2A ⊣ PKC

AMPA receptor subunits (Smith-Hicks *et al.*, 2010; Mikuni *et al.*, 2013; Kawata *et al.*, 2014). Together, all examples summarized in Table 1.1 show that the molecular machineries involved in LTD and synaptic pruning are largely overlapping, suggesting that they represent related cellular processes. A caveat in this analysis is that I have compared LTD and synaptic pruning at different types of synapses: PF synapses (LTD) and CF synapses (synaptic pruning). However, LTD has also been described at CF synapses in P14-30 rats (Hansel and Linden, 2000; Carta *et al.*, 2006). CF-LTD results from brief CF activation at 5 Hz and is, just like PF-LTD, postsynaptically induced and expressed (Shen *et al.*, 2002). CF-LTD has not been characterized in as much molecular detail as its counterpart at PF synapses. Even so, at Johns Hopkins University, David Linden and I were able to show that it requires a rise in calcium transients for its induction as well as the activation of mGluR1 receptors and PKC (Hansel and Linden, 2000). Thus, the available data suggest that the same mGluR1-PKC signaling cascade that triggers PF-LTD also induces CF-LTD, providing the missing link in the comparison of molecular pathways in LTD (at PF synapses) and synaptic pruning (at CF synapses).

## Synapse Elimination and LTD at the NMJ

At synapses between motor neurons and muscle fibers (NMJ), a developmental pruning process takes place that strongly resembles the elimination of surplus CF inputs in the cerebellum. At birth, each muscle fiber is innervated by approximately 10 motor neurons (Tapia *et al.*, 2012), which are subsequently eliminated until only one winner input remains at ~P16–18 (Redfern, 1970). As LTD-like depression has been observed at the developing (Lo and Poo, 1991) and adult NMJ (Etherington and Everett, 2004), we can address the question of how findings at NMJ synapses compare to the findings in the cerebellum. In the research field of NMJ plasticity, the focus has not been on linking molecular pathways to morphological pruning (as I described for cerebellar CF synapses), but rather on the physiological process of synaptic depression itself. For this reason, we have to look for similarities to LTD at other synapses, as well as the overlap in cellular machineries across ages. Synaptic depression at the developing NMJ is induced by repetitive depolarization and can be blocked by the calcium chelator BAPTA (Lo *et al.*, 1994). Photolytic calcium uncaging triggers LTD at these synapses as well (Cash *et al.*, 1996). These findings suggest that LTD at the NMJ depends on postsynaptic depolarization and subsequent calcium elevation. Moreover, depression at the developing NMJ requires CaMKII activation, whereas a form of potentiation depends on the activation of protein phosphatase 2B (calcineurin; Wan and Poo, 1999), pointing toward similar rules governing LTD and bidirectional synaptic plasticity as found in the cerebellum (Jörntell and Hansel, 2006). There is also evidence for molecular continuity: synaptic depression requires nitric oxide signaling at the developing (Wang *et al.*, 1995) and mature NMJ (Etherington and Everett, 2004). These findings support the notion that there is an LTD-like process at the developing NMJ that resembles LTD at the adult NMJ and marks synaptic inputs for elimination. A competition between multiple synaptic inputs is clearly demonstrated by the finding that laser removal of the stronger synaptic inputs rescues synapses that would have been eliminated otherwise (Turney and Lichtman, 2012). The same general finding has been made upon the ablation of the emerging winner CF input in the cerebellum

(Carillo *et al.*, 2013), again highlighting the similarity between the pruning processes at these two types of synapses.

## Temporal Continuity Between LTD-like Depression and Synapse Elimination

We have so far considered the argument of *molecular continuity*: synaptic pruning at developing synapses and LTD at adult synapses share large parts of their molecular machineries. But what about *temporal continuity*? The molecular overlap in LTD and pruning pathways makes it possible that synaptic disconnection may follow LTD. Although scarce, evidence for such temporal continuity does indeed exist. At the NMJ, synaptic weakening precedes disconnection (Colman *et al.*, 1997). Similarly, LTD at hippocampal synapses can be followed by synapse elimination (Nägerl *et al.*, 2004; Wiegert and Oertner, 2013). At cerebellar CF synapses, weak CF inputs undergo LTD, and it is the weaker synapses that are disconnected (Bosman *et al.*, 2008; Ohtsuki and Hirano, 2008). While these observations are made at developing synapses or synapses in cultures that remain in a juvenile state, the same phenomenon has been found at adult synapses, too. In the adult cerebellum, PF-LTD can be followed by the elimination of PF synapses that were depressed in a motor learning task (Wang *et al.*, 2014). At adult synapses, it remains to be seen whether synapse elimination represents a critical component of connectivity changes in learning. However, all these examples demonstrate that an LTD-like process marks synapses for elimination. A scenario emerges, in which a repertoire of plasticity mechanisms can serve functions at different stages of a brain's life with different but potentially overlapping meaning for the alteration of synaptic connectivity within local circuits over time.

## Termination of the Critical Period

The end of the critical period constitutes a caesura in brain plasticity. Nevertheless, the observation that there is molecular continuity in pathways used for pruning and adult plasticity leads to the question about what factors determine the duration of critical periods, and thus what factors terminate them. Critical periods mark a defined transition from high-amplitude plasticity during childhood development to subtle synaptic weight adjustments during adult learning. This transition is related to the maturation of a locally acting network of inhibitory interneurons that prevents large-scale alterations and thus catastrophic impact of recurring signals on an already matured brain (Hensch, 2005). The maturation of a subset of excitatory synapses that remain functionally disconnected for some time might have similar consequences for the termination of the critical period (Huang *et al.*, 2015). Thus, it seems likely that the duration of the critical period is related to the maturation of the excitatory and inhibitory synaptic network itself, rather than a switch in plasticity factors within neurons. This scenario aligns with the aspect of molecular continuity in plasticity pathways and an 'externally' enforced alteration in the magnitude of plasticity effects.

## A Role in Autism

Why is the molecular and temporal continuity of mechanisms underlying developmental synaptic pruning, along with LTD in the adult brain, so critical? First, the realization that developmental plasticity and pruning share components with a learning process, and that adult plasticity shares components with a maturation process enables a fresh look at the self-organization of neural circuits during a lifespan. In this new perspective, the brain never stops developing and optimizing even if plasticity outside the critical period is limited in magnitude. Development then becomes a learning process, further shaping brain circuits that were initially built following a raw genetic blueprint. Second, a recent focus on synaptopathies in developmental brain disorders (Grant, 2012; Zoghbi and Bear, 2012) has piqued new interest in synaptic dysfunction in autism spectrum disorder (ASD), particularly in deficits of synapse maturation and pruning. Brain overgrowth is one of the earliest signs in some forms of ASD (Courchesne *et al.*, 2003). Both brain hyperconnectivity (Supekar *et al.*, 2013) and hypoconnectivity (Mostofsky *et al.*, 2009; Di Martino *et al.*, 2014) have been found in children with autism. Interestingly, in ASD mouse models, deregulation of LTD is the most reliably found synaptic abnormality across brain areas and genetic types of autism (Piochon *et al.*, 2016). Understanding alterations in brain plasticity, particularly in pruning and LTD, might therefore hold one of the keys for deciphering which factors hamper proper mental development in autism.

# References

Aiba, A., Kano, M., Chen, C., Stanton, M.E., Fox, G.D., Herrup, K., Zwingman, T.A., and Tonegawa, S. (1994). Deficient cerebellar long-term depression and impaired motor learning in mGluR1 mutant mice. *Cell* 79, 377–388.

Bear, M.F., Kleinschmidt, A., Gu, Q.A., and Singer, W. (1990). Disruption of experience-dependent synaptic modifications in striate cortex by infusion of an NMDA receptor antagonist. *J. Neurosci.* 10, 909–925.

Bear, M.F., and Rittenhouse, C.D. (1999). Molecular basis for induction of ocular dominance plasticity. *J. Neurobiol.* 41, 83–91.

Belmeguenai, A., Botta, P., Weber, J.T., Carta, M., De Ruiter, M., De Zeeuw, C.I., Valenzuela, C.F., and Hansel, C. (2008). Alcohol impairs long-term depression at the cerebellar parallel fiber-Purkinje cell synapse. *J. Neurophysiol.* 100, 3167–3174.

Bienenstock, E.L., Cooper, L.N., and Munro, P.W. (1982). Theory for the development of neuron selectivity: orientation specificity and binocular interaction in visual cortex. *J. Neurosci.* 2, 32–48.

Bosman, L.W., Takechi, H., Hartmann, J., Eilers, J., and Konnerth, A. (2008). Homosynaptic long-term synaptic potentiation of the "winner" climbing fiber synapse in developing Purkinje cells. *J. Neurosci.* 28, 798–807.

Carillo, J., Nishiyama, N., and Nishiyama, H. (2013). Dendritic translocation establishes the winner in cerebellar climbing fiber elimination. *J. Neurosci.* 33, 7641–7653.

Carta, M., Mameli, M., and Valenzuela, C.F. (2006). Alcohol potently modulates climbing fiber—Purkinje neuron synapses: role of metabotropic glutamate receptors. *J. Neurosci.* 26, 1906–1912.

Cash, S., Dan, Y., Poo, M.M., and Zucker, R. (1996). Postsynaptic elevation of calcium induces persistent depression of developing neuromuscular synapses. *Neuron* 16, 745–754.

Chen, C., Kano, M., Chen, L., Bao, S., Kim, J.J., Hashimoto, K., Thompson, R.F., and Tonegawa, S. (1995). Impaired motor coordination correlates with persistent multiple climbing fiber innervation in PKCγ mutant mice. *Cell* 83, 1233–1242.

Chu, Y., Jin, X., Parada, I., Pesic, A., Stevens, B., Barres, B., and Prince, D.A. (2010). Enhanced synaptic connectivity and epilepsy in C1q knockout mice. *Proc. Natl. Acad. Sci. USA* 107, 7975–7980.

Chung, H.J., Steinberg, J.P., Huganir, R.L., and Linden, D.J. (2003). Requirement of AMPA receptor GluR2 phosphorylation for cerebellar long-term depression. *Science* 300, 1751–1755.

Colman, H., Nabekura, J., and Lichtman, J.W. (1997). Alterations in synaptic strength preceding axon withdrawal. *Science* 275, 356–361.

Conquet, F., Bashir, Z.I., Davies, C.H., Daniel, H., Ferraguti, F., Bordi, F., Franz-Bacon, K., Reggiani, A., Matarese, V., Conde, F., Collingridge, G.L., and Crepel, F. (1994). Motor deficit and impairment of synaptic plasticity in mice lacking mGluR1. *Nature* 372, 237–242.

Courchesne, E., Carper, R., and Akshoomoff, N. (2003). Evidence of brain overgrowth in the first year of life in autism. *J. Am. Med. Assoc.* 290, 337–344.

Crepel, F., Mariani, J., and Delhaye-Bouchaud, N. (1976). Evidence for a multiple innervation of Purkinje cells by climbing fibers in the immature rat cerebellum. *J. Neurobiol.* 7, 567–578.

De Zeeuw, C.I., Hansel, C., Bian, F., Koekkoek, S.K., van Alphen, A.M., Linden, D.J., and Oberdick, J. (1998). Expression of a protein kinase C inhibitor in Purkinje cells blocks cerebellar LTD and adaptation of the vestibulo-ocular reflex. *Neuron* 20, 495–508.

Di Martino, A., *et al.* (2014). The autism brain imaging data exchange: towards a large-scale evaluation of the intrinsic brain architecture in autism. *Mol. Psychiatry* 19, 659–667.

Etherington, S.J., and Everett, A.W. (2004). Postsynaptic production of nitric oxide implicated in long-term depression at the mature amphibian (*Bufo marinus*) neuromuscular junction. *J. Physiol.* 559, 507–517.

Grant, S.G. (2012). Synaptopathies: diseases of the synaptome. *Curr. Opin. Neurobiol.* 22, 522–529.

Hansel, C., and Linden, D.J. (2000). Long-term depression of the cerebellar climbing fiber-Purkinje neuron synapse. *Neuron* 26, 473–482.

Hansel, C., de Jeu, M., Belmeguenai, A., Houtman, S.H., Buitendijk, G.H., Andreev, D., De Zeeuw, C.I., and Elgersma, Y. (2006). αCaMKII is essential for cerebellar LTD and motor learning. *Neuron* 51, 835–843.

Hartmann, J., Blum, R., Kovalchuk, Y., Adelsberger, H., Kuner, R., Durand, G.M., Miyata, M., Kano, M., Offermanns, S., and Konnerth, A. (2004).

Distinct roles of Galpha(q) and Galpha11 for Purkinje cell signaling and motor behavior. *J. Neurosci.* 24, 5119–5130.

Hashimoto, K., and Kano, M. (2003). Functional differentiation of multiple climbing fiber inputs during synapse elimination in the developing cerebellum. *Neuron* 38, 785–796.

Hashimoto, K., and Kano, M. (2013). Synapse elimination in the developing cerebellum. *Cell. Mol. Life Sci.* 70, 4667–4680.

Hashimoto, K., Ichikawa, R., Kitamura, K., Watanabe, M., and Kano, M. (2009). Translocation of a "winner" climbing fiber to the Purkinje cell dendrite and subsequent elimination of "losers" from the soma in developing cerebellum. *Neuron* 63, 106–118.

Hashimoto, K., Miyata, M., Watanabe, M., and Kano, M. (2001). Roles of phospholipase Cbeta4 in synapse elimination and plasticity in developing and mature cerebellum. *Mol. Neurobiol.* 23, 69–82.

Hashimoto, K., Tsujita, M., Miyazaki, T., Kitamura, K., Yamazaki, M., Shin, H.S., Watanabe, M., Sakimura, K., and Kano, M. (2011). Postsynaptic P/Q-type Ca2+ channel in Purkinje cell mediates synaptic competition and elimination in developing cerebellum. *Proc. Natl. Acad. Sci. USA* 108, 9987–9992.

Hashimoto, K., Yoshida, T., Sakimura, K., Mishina, M., Watanabe, M., and Kano, M. (2009). Influence of parallel fiber-Purkinje cell synapse formation on postnatal development of climbing fiber-Purkinje cell synapses in the cerebellum. *Neuroscience* 162, 601–611.

Hensch, T.K. (2005). Critical period plasticity in local cortical circuits. *Nat. Rev. Neurosci.* 6, 877–888.

Heynen, A.J., Yoon, B.J., Liu, C.H., Chung, H.J., Huganir, R.L., and Bear, M.F. (2003). Molecular mechanism for loss of visual cortical responsiveness following brief monocular deprivation. *Nat. Neurosci.* 6, 854–862.

Hoshiko, M., Arnoux, I., Avignone, E., Yamamoto, N., and Audinat, E. (2012). Deficiency of the microglial receptor CX3CR1 impairs postnatal development of thalamocortical synapses in the barrel cortex. *J. Neurosci.* 32, 15106–15111.

Huang, X., Stodieck, S.K., Goetze, B., Cui, L., Wong, M.H., Wenzel, C., Hosang, L., Dong, Y., Löwel, S., and Schlüter, O.M. (2015). Progressive maturation of silent synapses governs the duration of a critical period. *Proc. Natl. Acad. Sci. USA* 112, 3131–3140.

Ichise, T., Kano, M., Hashimoto, K., Yanagihara, D., Nakao, K., Shigemoto, R., Katsuki, M., and Aiba, A. (2000). mGluR1 in cerebellar Purkinje cells essential for long-term depression, synapse elimination, and motor coordination. *Science* 288, 1832–1835.

Ito, M., Yamaguchi, K., Nagao, S., and Yamazaki, T. (2014). Long-term depression as a model of cerebellar plasticity. *Prog. Brain Res.* 210, 1–30.

Jörntell, H., and Hansel, C. (2006). Synaptic memories upside down: bidirectional plasticity at cerebellar parallel fiber-Purkinje cell synapses. *Neuron* 52, 227–238.

Kano, M., Hashimoto, K., Chen, C., Abeliovich, A., Aiba, A., Kurihara, H., Watanabe, M., Inoue, Y.L., and Tonegawa, S. (1995). Impaired synapse elimination during cerebellar development in PKCγ mutant mice. *Cell* 83, 1223–1231.

Kano, M., Hashimoto, K., Kurihara, H., Watanabe, M., Inoue, Y., Aiba, A., and Tonegawa, S. (1997). Persistent multiple climbing fiber innervation of cerebellar Purkinje cells in mice lacking mGluR1. *Neuron* 18, 71–79.

Kano, M., Hashimoto, K., Watanabe, M., Kurihara, H., Offermanns, S., Jiang, H., Wu, Y., Jun, K., Shin, H.S., Inoue, Y., Simon, M.I., and Wu, D. (1998). Phospholipase cbeta4 is specifically involved in climbing fiber synapse elimination in the developing cerebellum. *Proc. Natl. Acad. Sci. USA* 95, 15724–15729.

Kawamura, Y., Nakayama, H., Hashimoto, K., Sakimura, K., Kitamura, K., and Kano, M. (2013). Spike-timing-dependent selective strengthening of single climbing fibre inputs to Purkinje cells during cerebellar development. *Nat. Commun.* 4, 2732.

Kawata, S., Miyazaki, T., Yamazaki, M., Mikuni, T., Yamasaki, M., Hashimoto, K., Watanabe, M., Sakimura, K., and Kano, M. (2014). Global scaling down of excitatory postsynaptic responses in cerebellar Purkinje cells impairs developmental synapse elimination. *Cell Rep.* 8, 1119–1129.

Kirkwood, A., and Bear, M.F. (1994). Homosynaptic long-term depression in the visual cortex. *J. Neurosci.* 14, 3404–3412.

Kirkwood, A., Dudek, S.M., Gold, J.T., Aizenman, C.D., and Bear, M.F. (1993). Common forms of synaptic plasticity in the hippocampus and neocortex in vitro. *Science* 260, 1518–1521.

Kleinschmidt, A., Bear, M.F., and Singer, W. (1987). Blockade of "NMDA" receptors disrupts experience-dependent plasticity of kitten striate cortex. *Science* 238, 355–358.

Knöpfel, T., and Grandes, P. (2002). Metabotropic glutamate receptors in the cerebellum with a focus on their function in Purkinje cells. *Cerebellum* 1, 19–26.

Leitges, M., Kovac, J., Plomann, M., and Linden, D.J. (2004). A unique PDZ ligand in PKCalpha confers induction of cerebellar long-term synaptic depression. *Neuron* 44, 585–594.

Levenes, C., Daniel, H., Jaillard, D., Conquet, F., and Crepel, F. (1997). Incomplete regression of multiple climbing fibre innervation of cerebellar Purkinje cells in mGluR1 mutant mice. *NeuroReport* 8, 571–574.

Linden, D.J., and Connor, J.A. (1991). Participation of postsynaptic PKC in cerebellar long-term depression in culture. *Science* 254, 1656–1659.

Lo, Y.J., and Poo, M.M. (1991). Activity-dependent synaptic competition in vitro: heterosynaptic depression of developing synapses. *Science* 254, 1019–1022.

Lo, Y.J., Lin, Y.C., Sanes, D.H., and Poo, M.M. (1994). Depression of developing neuromuscular synapses induced by repetitive depolarizations. *J. Neurosci.* 14, 4694–4704.

Lohmann, C., and Kessels, H.W. (2014). The developmental stages of synaptic plasticity. *J. Physiol.* 592, 13–31.

Mikuni, T., Uesaka, N., Okuno, H., Hirai, H., Deisseroth, K., Bito, H., and Kano, M. (2013). Arc/Arg 3.1 is a postsynaptic mediator of activity-dependent synapse elimination in the developing cerebellum. *Neuron* 78, 1024–1035.

Mostofsky, S.H., Powell, S.K., Simmonds, D.J., Goldberg, M.C., Caffo, B., and Pekar, J.J. (2009). Decreased connectivity and cerebellar activity in autism during motor task performance. *Brain* 132, 2413–2425.

Nägerl, U.V., Eberhorn, N., Cambridge, S.B., and Bonhoeffer, T. (2004). Bidirectional activity-dependent morphological plasticity in hippocampal neurons. *Neuron* 44, 759–767.

Offermanns, S., Hashimoto, K., Watanabe, M., Sun, W., Kurihara, H., Thompson, R.F., Inoue, Y., Kano, M., and Simon, M.I. (1997). Impaired motor coordination and persistent multiple climbing fiber innervation of cerebellar Purkinje cells lacking Galphaq. *Proc. Natl. Acad. Sci. USA* 94, 14089–14094.

Ohtsuki, G., and Hirano, T. (2008). Bidirectional plasticity at developing climbing fiber-Purkinje neuron synapses. *Eur. J. Neurosci.* 28, 2393–2400.

Paolicelli, R.C., Bolasco, G., Pagani, F., Maggi, L., Scianni, M., Panzanelli, P., Giustetto, M., Ferreira, T.A., Guiducci, E., Dumas, L., Ragozzino, D., and Gross, C.T. (2011). Synaptic pruning by microglia is necessary for normal brain development. *Science* 333, 1456–1458.

Piochon, C., Irinopoulo, T., Brusciano, D., Bailly, Y., Mariani, J., and Levenes, C. (2007). NMDA receptor contribution to the climbing fiber response in the adult mouse Purkinje cell. *J. Neurosci.* 27, 10797–10809.

Piochon, C., Kano, M., and Hansel, C. (2016). LTD-like molecular pathways in developmental synaptic pruning. *Nat. Neurosci.* 19, 1299–1310.

Piochon, C., Levenes, C., Ohtsuki, G., and Hansel, C. (2010). Purkinje cell NMDA receptors assume a key role in synaptic gain control in the mature cerebellum. *J. Neurosci.* 30, 15330–15335.

Redfern, P.A. (1970). Neuromuscular transmission in new-born rats. *J. Physiol.* 209, 701–709.

Renzi, M., Farrant, M., and Cull-Candy, S.G. (2007). Climbing-fibre activation of NMDA receptors in Purkinje cells of adult mice. *J. Physiol.* 585, 91–101.

Rittenhouse, C.D., Shouval, H.Z., Paradiso, M.A., and Bear, M.F. (1999). Monocular deprivation induces homosynaptic long-term depression in visual cortex. *Nature* 397, 347–350.

Shen, Y., Hansel, C., and Linden, D.J. (2002). Glutamate release during LTD at cerebellar climbing fiber-Purkinje cell synapses. *Nat. Neurosci.* 5, 725–726.

Sidorov, M.S., Kaplan, E.S., Osterweil, E.K., Lindemann, L., and Bear, M.F. (2015). Metabotropic glutamate receptor signaling is required for NMDA receptor-dependent ocular dominance plasticity and LTD in visual cortex. *Proc. Natl. Acad. Sci. USA* 112, 12852–12857.

Singer, W. (1995). Development and plasticity of cortical processing architectures. *Science* 270, 758–764.

Smith-Hicks, C., Xiao, B., Deng, R., Ji, Y., Zhao, X., Shepherd, J.D., Posern, G., Kuhl, D., Huganir, R.L., Ginty, D.D., Worley, P.F., and Linden, D.J. (2010). SRF binding to SRE 6.9 in the Arc promoter is essential for LTD in cultured Purkinje cells. *Nat. Neurosci.* 13, 1082–1089.

Supekar, K., Uddin, L.Q., Khouzam, A., Phillips, J., Gaillard, W.D., Kenworthy, L.E., Yerys, B.E., Vaidya, C.J., and Menon, V. (2013). Brain hyperconnectivity in children with autism and its links to social deficits. *Cell Rep.* 5, 738–747.

Tapia, J.C., Wylie, J.D., Kasthuri, N., Hayworth, K.J., Schalek, R., Berger, D.R., Guatimosim, C., Seung, H.S., and Lichtman, J.W. (2012). Pervasive synaptic branch removal in the mammalian neuromuscular system at birth. *Neuron* 74, 816–829.

Turney, S.G., and Lichtman, J.W. (2012). Reversing the outcome of synapse elimination at developing neuromuscular junctions in vivo: evidence for synaptic competition and its mechanism. *PLOS Biology* 10, e1001352.

Wan, J., and Poo, M.M. (1999). Activity-induced potentiation of developing neuromuscular synapses. *Science* 285, 1725–1728.

Wang, T., Xie, Z., and Lu, B. (1995). Nitric oxide mediates activity-dependent synaptic suppression at developing neuromuscular synapses. *Nature* 374, 262–266.

Wang, W., Nakadate, K., Masugi-Tokita, M., Shutoh, F., Aziz, W., Tarusawa, E., Lorincz, A., Molnar, E., Kesaf, S., Li, Y.Q., Fukazawa, Y., Nagao, S., and Shigemoto, R. (2014). Distinct cerebellar engrams in short-term and long-term motor learning. *Proc. Natl. Acad. Sci. USA* 111, E188–E193.

Wiegert, J.S., and Oertner, T.G. (2013). Long-term depression triggers the selective elimination of weakly integrated synapses. *Proc. Natl. Acad. Sci. USA* 110, E4510–E4519.

Wiesel, T.N., and Hubel, D.H. (1963). Effects of visual deprivation on the morphology and physiology of cells in the cat's lateral geniculate body. *J. Neurophysiol.* 26, 978–993.

Zhan, Y., Paolicelli, R.C., Sforazzini, F., Weinhard, L., Bolasco, G., Pagani, F., Vyssotski, A.L., Bifone, A., Gozzi, A., Ragozzino, D., and Gross, C.T. (2014). Deficient neuron-microglia signaling results in impaired functional brain connectivity and social behavior. *Nat. Neurosci.* 17, 400–406.

Zoghbi, H.Y., and Bear, M.F. (2012). Synaptic dysfunction in neurodevelopmental disorders associated with autism and intellectual disabilities. *Cold Spring Harb. Perspect. Biol.* 4, a009886.

# 5 Synaptopathies: Synaptic Dysfunction in Autism

## Genetic and Environmental Factors Contribute to Autism

Brain developmental disorders, including autism, can arise from genetic and environmental causes that may alter morphological connectivity patterns in the brain. For example, the corpus callosum—the 'bridge' between the cortical hemispheres—shows differences in the volume of the total bridge structure, as well as the thickness of individual fibers within the bridge in the developing brain of autistic children (e.g. Wolff *et al.*, 2015; Fingher *et al.*, 2017). Structural abnormalities have also been found in other brain areas, some of which will be discussed later. An additional type of alteration can be found at the synapse, and this is the type of pathological change that I will focus upon here. Autism and other brain developmental disorders are, at least in part, considered diseases of the synapse.[1] An identification of abnormalities shared by all disorders 'on the spectrum' (Autism Spectrum, see the following) has not been possible, although in this chapter I will make an effort to highlight recurrent motifs of cellular alterations that are typical for some groups of disorders within the spectrum.

The clinical definition of autism and its related neurodevelopmental disorders was significantly updated with the publication of the fifth edition of the *Diagnostic and Statistical Manual of Mental*

---

[1] I am rephrasing the title of S.G. Grant's review paper 'Synaptopathies: diseases of the synaptome' (*Curr. Opin. Neurobiol.* 22, 522–529, 2012).

*Disorders* (DSM-5) by the American Psychiatric Association in 2013. Four previously separately diagnosed disorders—autistic disorder, Asperger's disorder, childhood disintegrative disorder, and pervasive developmental disorder not otherwise specified—were now merged into 'Autism Spectrum Disorder' (ASD). This merge was intended to create a wider autism-related umbrella term to reduce diagnostic errors and inconsistencies, and to make sure that specific subgroups of children—for example, those with mild cognitive impairment (as in Asperger's syndrome)—would not be overlooked.[2] The practical consequences of these changes for clinicians and affected families and individuals need to be evaluated elsewhere. From a scientific perspective, the loss of differentiation is problematic. Asperger's patients have a normal to high IQ and little to no impairment of language, whereas other syndromic forms of autism ('syndromic' means that autism is one of several types of symptoms) are often characterized by low IQ and strong impairment or even absence of language. Both Asperger's syndrome and these other syndromic forms of autism fall under the diagnosis of ASD, but have potentially more symptomatic differences than similarities, and will likely not share the same causes. An ASD diagnosis includes these criteria:

— *Persistent deficits in social communication and interaction across multiple contexts*
— *Restricted, repetitive patterns of behavior, interests, or activities*

Both criteria need to be met for the ASD diagnosis. Individually, they might also be found in other diseases that show general symptomatic overlap with autism, such as anxiety disorders or obsessive-compulsive disorder (OCD). In addition, ASD may be accompanied by other symptoms such as altered sensitivity to sensory input, for example to touch or noise, and motor problems. I will discuss these ASD-related symptoms later in this chapter and in Chapter VI. But

---

[2] However, the concern was also raised that the opposite might happen: that lumping Asperger's syndrome together with autism would lead to the loss of specific attention and support.

take note that their association with autism is not reliable enough to be included into the core diagnostic criteria specified in DSM-5. (Hyper-or hyposensitivity to sensory input, or an unusual interest in specific types of sensory input can be one of two required observations to confirm the presence of 'restricted, repetitive behaviors', though.)

The lack of differentiation in the diagnosis of ASD not only results from overarching goals in health policies, but also from this large overlap in symptoms. Perhaps with the exception of Asperger's Syndrome, there are no ASD subgroups that could be easily distinguished by specific, defined symptoms. With only slight exaggeration, one could make the provocative statement that each individual ASD patient has their own disorder, different from that of the next, as much as one could say that all have the same disorder, but with a multitude of different symptomatic expressions. The resulting lack of precision in the description of ASD makes it even harder to identify developmental abnormalities in the brain that cause autism. It is known that genetic and environmental factors can cause ASD or contribute to it. Of note, there is no scientific evidence supporting the still widespread fear that vaccination against measles, mumps, and rubella (MMR) causes autism and thus represents one of these environmental factors. The original 1998 study that made such claim has since been retracted by the *Lancet*, the journal that originally published it,[3] after a ruling by the UK General Medical Council that '......several elements of the 1998 paper.....are incorrect...'. The current consensus among mental health experts is well summarized in a statement by the Mayo Clinic published on their website: 'Vaccines do not cause autism. Despite much controversy on the topic, researchers haven't found a connection between autism and childhood vaccines'.[4] Parents look for explanations about why their child is autistic. Unsubstantiated claims, such as the vaccine theory, hurt affected families as they suggest simple solutions ('stop MMR

[3] The Editors of The Lancet (2010): Retraction—Ileal-lymphoid-nodular hyperplasia, non-specific colitis, and pervasive developmental disorder in children. doi:10.1016/S0140-6736(10)60175-4.
[4] www.mayoclinic.org/healthy-lifestyle/infant-and-toddler-health/in-depth/vaccines/

vaccination') that will not lead to the desired outcome. But what do we know about the causes of autism? Environmental factors that indeed seem to promote autism include chemicals that, during pregnancies, were once widely used as sedatives (thalidomide) or against migraines or seizures (valproic acid) (Landrigan, 2010; Ornoy *et al.*, 2015). While valproic acid is still used to reduce seizures during pregnancies (it is not prescribed any longer to treat migraines in pregnant women), thalidomide was taken off the market in 1961, after it was discovered that it causes severe limb malformations in infants. In Germany, where the drug was developed and sold since 1957 by Chemie-Grünenthal under the trade name Contergan, up to 7000 children were born with these malformations.[5] It is now believed that the limb malformations resulted from the antiangiogenic actions (inhibition of blood vessel growth) of thalidomide (D'Amato *et al.*, 1994). The autism-promoting effects of thalidomide and valproic acid might be related to their modulatory effects on genes involved in proliferation, neuronal differentiation, and synaptogenesis (Dufour-Rainfray *et al.*, 2011), providing an example of how environmental influences can lead to autism through their impact on gene regulation.

Next to environmental factors that might enhance the risk of autism—most of which are still unidentified—there is an obvious and better investigated genetic component as well. Autism is highly heritable. When one identical twin has autism, the chance that the other twin has autism, too, is about 80% (Colvert *et al.*, 2015). The New York-based Simons Foundation Autism Research Initiative (SFARI), the largest private funder of autism research in the world, reports that of the 19,000–20,000 protein coding genes in the human genome, 1,089 genes have been implicated in autism (as of 2019); 91 of these are high-ranking autism risk genes.[6] Many of these genes code for proteins involved either in the regulation of gene expression or in synaptic function (Sestan and State, 2018; Figure 5.1).

---

[5]An insightful article on the Contergan scandal has been published in 2014 by Harold Evans in *The Guardian*: 'Thalidomide: how men who blighted lives of thousands evaded justice'.
[6]www.sfari.org: SFARI Resources/SFARI Gene.

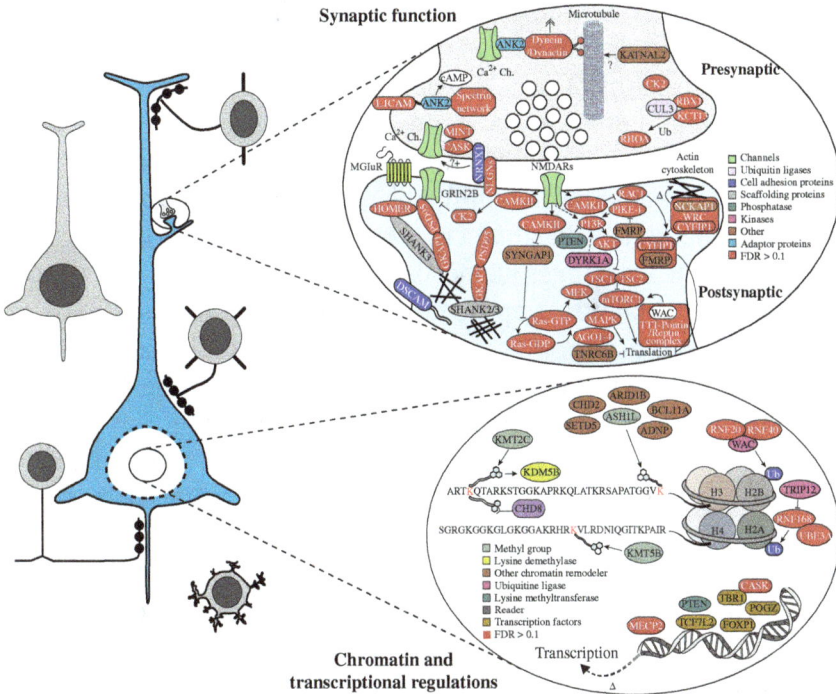

**Figure 5.1:** High-risk ASD genes encode synaptic proteins and transcriptional regulators. The scheme depicts proteins encoded by high-risk ASD genes that were identified by statistical analysis of databases. Left: neuron with other cells within a neural circuit. Right: schemes illustrating signaling factors in gene regulation (bottom) and synaptic function (top) that are likely involved in autism. The figure is taken from Sestan and State (2018).

The dominance of genes involved in the regulation of other genes, and of genes involved in synaptic function, is no surprise. The former, because autism is a disorder of brain development, which is tightly controlled by gene regulatory networks. The latter, because autism is a cognitive disorder that impairs the communication between neurons. Analysis of ASD genomics has significantly advanced the field over recent years, and even constitutes a major research approach that is available to study autism directly in populations of human patients. That said, one limitation to human studies is that they offer no insight into the alterations within the autistic brain, such as the cellular abnormalities that alter synaptic function and neural circuits. To be fair, measurements of brain activity in humans, for example, using functional magnetic resonance imaging (fMRI) and high-density electroencephalogram (EEG) recordings, have significantly contributed to our understanding of altered connectivity within and between specific brain areas. However, these methods do not have the spatial resolution that is required to examine synapses, individual neurons, and small circuits—in short, the small structures where the critical changes likely take place. Our ultimate goal should be to obtain an (almost) gap-free understanding of how ASD-related changes at an initial level of deregulation (genes and/or environmental factors) affect the next higher level, synaptic connectivity, and function, and how these synaptic changes cause dysfunction of larger populations of neurons (circuits). These discoveries will eventually lead to a plausible explanation of altered behavioral repertoires (Figure 5.2).

The suggestion to identify genotype–phenotype[7] event sequences, such as the one illustrated here, implies that there is true causation along a mono-factorial axis. This is possibly an over-simplification. It is known that human traits, such as body height (Wood *et al.*, 2014), IQ (Davies *et al.*, 2011), and sexual orientation (Ganna *et al.*, 2019) result from complex interactions between multiple genes and environmental factors, such as receiving maternal hormones during

**Figure 5.2:** Four levels of investigation of abnormalities in autism. Genetic alterations, which can include point mutations and chromosome defects, will alter synaptic function and connectivity. As a result, local neural circuits generate different output signals. These alterations will have an effect on behavior of a human with autism, or a model organism, for example, on social interaction behaviors. 'Behavior' here also stands for other consequences of altered brain function, such as the enhanced occurrence of seizures that is typical for some forms of autism.

---

[7]The term 'genotype' describes the complete set of genes of an organism. The term 'phenotype' (related to 'phenomenon'; from the Greek verb φαίνειν, phainein, which means to appear, or to manifest itself) describes observable physical and behavioral characteristics that in part result from the genotype.

pregnancy. A similar complexity of contributing factors is likely to exist in many cases of autism, when the specific genomic landscape—potentially together with non-genetic factors—creates a permissive environment for the expression of the autism phenotype. However, autism can also be caused by isolated genetic defects. Nobel Prize winner Thomas Südhof from Stanford University demonstrated that in otherwise normal mice introduction of a specific point mutation in the gene coding for neuroligin-3, a cell adhesion molecule required for the proper organization of synapses, causes deficits in social interactions. This point mutation was identified in a small number of ASD patients in a Swedish family (Tabuchi *et al.*, 2007). This study provides a compelling proof-of-principle example, showing that a point mutation in a specific synaptic protein can lead to the autism phenotype. Subsequent studies have linked specific de novo mutations in humans to an autism outcome (documented, e.g. by the Simons Simplex Collection), providing further support for the notion that isolated gene defects can lead to autism. In fact, the introduction of ASD candidate mutations into mice enables a search for synaptic and/or circuit alterations caused by this genetic manipulation. That said, one challenge with this type of study is the validation of the mouse model—that is, the demonstration that these mice show autism-resembling behaviors. As I will discuss in Chapter VI, tests for social interaction deficits in mice do not readily reflect human behavior alterations, and only gain value when performed with carefully selected control experiments. Further, they must be complemented with tests for more directly comparable behaviors, such as motor control and learning, or evasive/protective behaviors that reflect sensitivity to noxious stimuli.

# Cell Physiological Signatures of the Autistic Brain[8]

The complexity of cognitive behaviors currently prevents a comprehensive analysis of which synapse and circuit functions, in their

---

[8]This paragraph is in parts adapted from Piochon *et al.* (2016) and Hansel (2019).

entirety, drive cognition. And further, as a logical consequence, which alterations of such functions contribute to brain developmental disorders? However, it is possible to identify specific motifs that are potentially significant. There are two established observations on abnormalities in the structure and function of neurons in autistic brains: (a) reduced pruning of synapses and spines (Hutsler and Zhang, 2010; Tang *et al.*, 2014; Wang *et al.*, 2017) and (b) an enhanced ratio of excitatory to inhibitory synaptic input, along with enhanced overall excitability of neurons. The latter phenomenon is likely to underlie the increase in sensitivity to sensory input in some autistic children, leading to the description of autism as the 'Intense World Syndrome' (Markram *et al.*, 2007).

A surprising recurrent finding in mouse models of autism— across different genetic defects and across different brain areas—is the deregulation of LTD (Table 5.1). This type of synaptic dysfunction has been found in mouse models for Fragile X syndrome (*Fmr1* knockout; Huber *et al.*, 2002; Koekkoek *et al.*, 2005), as well as mouse models for Tuberous Sclerosis (*Tsc2*[+/-], Auerbach *et al.*, 2011). Moreover, impaired LTD has been found in a mouse model for the human 15q11-13 duplication (Dup15q syndrome, patDp/+ mice; Piochon *et al.*, 2014). While these are all examples of syndromic autism, LTD deregulation has also been found in nonsyndromic autism, in neuroligin-3 knockout mice (Baudouin *et al.*, 2012),[9] and even with prenatal exposure to valproic acid in rats (Zhang *et al.*, 2003). This wide range of different ASD factors demonstrates that LTD deregulation might be a common feature in autism. Interestingly, in all cases listed, LTD is enhanced or saturated, with the exception of *Tsc2*[+/-] mice (Auerbach *et al.*, 2011) and patDp/+ mice (Piochon *et al.*, 2014), where LTD is reduced or prevented. The observation that the type of LTD deregulation might fall on different ends of plasticity changes does not, however, exclude similar consequences. This is because LTD saturation limits the dynamic range of plasticity and might prevent subsequent synaptic depression under appropriate physiological conditions.

---

[9]LTD is, however, intact in neuroligin 1, 2, 3 triple knockouts (Zhang *et al.*, 2015). The reason for this discrepancy is currently unknown.

**Table 5.1:** LTD deregulation has been observed in several mouse models of autism and across various brain areas. Developmental elimination of surplus climbing fibers (CFs) as a measure for synaptic pruning has been tested in all cerebellar studies shown. N/D, not determined. N/A, not applicable. This table is adapted from Piochon *et al.* (2016).

| Mouse | Brain area | LTD | CF elimination (age tested) | References |
|---|---|---|---|---|
| *Fmr1* knockout | hippocampus cerebellum | enhanced enhanced | N/A accelerated (P21–48) | Huber *et al.* (2002) Koekkoek *et al.* (2005) |
| *Nlgn3* knockout | cerebellum | impaired* | normal; (2–3 mo) | Baudouin *et al.* (2012) |
| Nlgn3-R451C | cerebellum | N/D | delayed (P11–17; >P17) | Lai *et al.* (2013) |
| Nl-1/2/3 knockout | cerebellum | normal | normal | Zhang *et al.* (2015) |
| *Syngap* +/– | hippocampus | enhanced | N/A | Barnes *et al.* (2015) |
| eIF4E-transgenic | hippocampus striatum | enhanced enhanced | N/A | Santini *et al.* (2013) |
| *Tsc2*+/– | hippocampus | reduced | N/A | Auerbach *et al.* (2014) |
| 15q11–13 duplication | cerebellum | impaired | delayed (P10–12; 2 mo) | Piochon *et al.* (2014) |

A close inspection of genetic aberrations in syndromic forms of autism reveals a remarkable convergence on signaling pathways involved in protein translation, the process of protein synthesis based on the genetic code (Figure 5.3). These signaling pathways make up a complex network that is regulated, among other factors, by a molecule called 'mTOR' (mammalian target of rapamycin). This molecule also plays a role in cell growth and proliferation, and is, perhaps not surprisingly, involved in tumor growth. mTOR signaling promotes the formation of a complex of factors that are intimately involved in the initiation of translation, eIF4E and eIF4G ('eIF' stands for eukaryotic initiation factor). In the brain, mTOR signaling can be triggered by the activation of group I metabotropic glutamate receptors (mGluRs; Baudouin, 2014; Santini and Klann, 2014; Huber *et al.*, 2015), and thus 'responds' to synaptic transmission. Indeed, group I mGluRs (mGluR1 and mGluR5) initiate protein translation that can

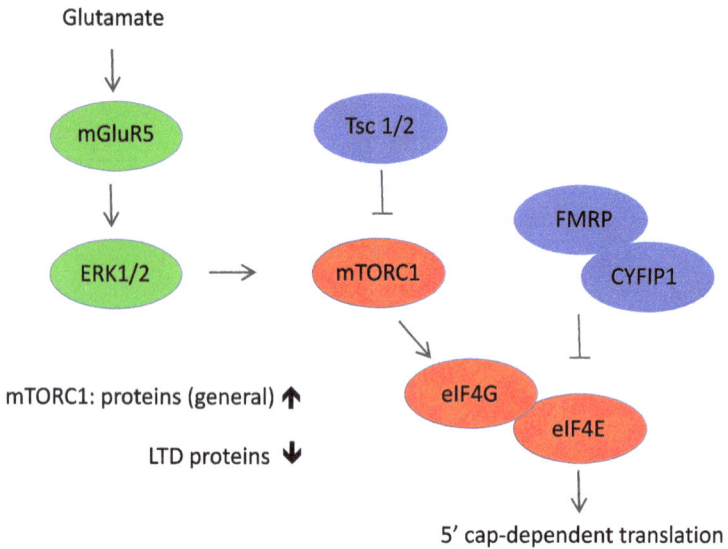

**Figure 5.3:** ASD candidate genes converge onto a network of translation control factors. Blue: proteins encoded by these ASD candidate genes. Red: factors involved in immediate control of translation (5' cap-dependent translation describes a subgroup of translation machineries controlled by the eIF4E/eIF4G initiation complex). Green: factors that trigger translation in response to glutamate signaling. mTOR generally promotes protein synthesis, but seems to suppress the synthesis of LTD proteins. This would explain why, as described in Auerbach *et al.* (2011), LTD is altered in opposite directions in mouse models for TSC and Fragile X syndrome, respectively.

occur locally in the dendrites (Weiler and Greenough, 1993). The molecular pathway triggered by group I mGluRs that upregulates translation is controlled by several proteins affected in forms of syndromic autism, including TSC1/2 and Fragile X Mental Retardation Protein (FMRP). The resulting critical role of metabotropic signaling in autism is highlighted in the 'mGluR theory of Fragile X syndrome' developed by Mark Bear and colleagues at MIT (Bear *et al.*, 2004; Osterweil *et al.*, 2010). In this theory, the lack of FMRP leads to excessive protein synthesis downstream of group I mGluR activation, including proteins that promote the induction of LTD. Together with its binding partner CYFIP1 (Cytoplasmic FMRP interacting protein 1), FMRP regulates the ability of the translation factor eIF4E to initiate translation (Darnell *et al.*, 2011; Santini and Klann, 2014). The gene

*CYFIP1* is located on the proximal end of chromosome 15q11–13 and is indeed upregulated in postmortem tissue of some Dup15q syndrome patients (Oguro-Ando *et al.*, 2015). Overexpression of eIF4E in transgenic mice causes ASD-like behavioral phenotypes, as well as enhanced LTD in the hippocampus and striatum (Santini *et al.*, 2013). These findings suggest that altered translation can cause synaptic abnormalities associated with autism, including LTD deregulation. Mutations affecting TSC1/2 and FMRP signaling seem to have opposing effects on the translation of synaptic proteins (incl. LTD-relevant), but both cause autism and intellectual disability (Auerbach *et al.*, 2011). Similarly, it has recently been shown that both up- and downregulation of mTOR signaling impair song learning in birds (Ahmadianteherani and London, 2017); this phenomenon shows resemblance to vocal communication in humans. These findings indicate that translation, particularly when controlling local, dendritic synthesis of synaptic proteins, needs to be properly regulated and balanced. Further, any and all deviations from this balance can have catastrophic consequences for synapse and circuit function. It is almost certain that LTD-regulating proteins only represent a small subgroup of important proteins that are affected. Even so LTD deregulation is consistently found in ASD models and is currently one of the most promising leads in ASD research.

## Autism Phenotypes that may Result from LTD Deregulation

The central position that altered genes and proteins found in some types of ASD assume in translation regulatory networks might explain the complex phenotypes that result. It also explains why syndromic autism is so prominently linked to these genetic defects. After all, translation control sits 'upstream' of the many other cellular processes, all of which depend on specific spatially and temporally controlled protein expression patterns. LTD deregulation is certainly only one of many cell pathological alterations that result from these genetic defects. And yet, an in-depth discussion of LTD deregulation is appropriate, because it is a neurophysiological problem that has been described in some detail, and because it might hold one of the keys to understand cognitive deficits in autism.

*LTD and synaptic pruning.* I have already discussed the close relationship between LTD and synaptic pruning. The large overlap in molecular machineries that steer both processes essentially means that LTD, or LTD-like processes, provide a plasticity component in the early stages of pruning dominated by competition between synaptic inputs. A reduction in pruning will not only lead to hyper-connectivity and hyper-excitability, but will also make it impossible for the brain to optimize connectivity, a process that is required to enable selective encoding of incoming information. Thus, the resulting network problem is not only a quantitative one (too many synapses) but also a qualitative one (the wrong synapses are maintained). This encoding problem might well contribute to cognitive deficits in autism, when neurons do not receive the right type of information from other neurons.

*LTD and motor coordination/learning.* About 80% of children with autism show motor impairment, including general clumsiness and problems with eye movement control (Green *et al.*, 2009; Fournier *et al.*, 2010; Johnson *et al.*, 2013; Mosconi *et al.*, 2013). These observations point toward the cerebellum, a brain area that is involved in the fine-tuning of movements (Fatemi *et al.*, 2012; Wang *et al.*, 2014; Mosconi *et al.*, 2015), playing a role in autism. This connection is further supported by the finding that ASD patients show abnormalities in classical eyeblink conditioning (Sears *et al.*, 1994; Oristaglio *et al.*, 2013; Welsh and Oristaglio, 2016), a laboratory test for motor learning, whose successful completion requires an intact cerebellum (McCormick and Thompson, 1984). In eyeblink conditioning, a so-called unconditioned stimulus (US) is presented, typically a brief air puff that is blown onto the cornea of the eye, which leads to a closure of the eyelid. This is a protective reflex. The learning takes place, when a previously neutral stimulus that on its own does not cause the eyelid closure, such as a tone or a light signal (the 'conditioned stimulus', CS), is repeatedly presented together with the US. In the 'delay' version of the test used here, the CS onset precedes

US onset by about 250 ms, and both coterminate. As a result of the conditioning period, which can be spaced in distinct sessions over several days, the CS alone initiates the eyelid closure, the 'conditioned response' (CR; Figure 5.4). In essence, eyeblink conditioning is a form of associative learning, in which an association is established between two stimuli, a neutral stimulus and a potentially harmful stimulus. In classic Albus-Ito theories of cerebellar function, LTD is a synaptic plasticity that mediates this form of learning. This is because of the fact that synaptic depression at the excitatory parallel fiber (PF) synapses onto Purkinje cells causes a reduction of Purkinje cell output. In turn, this reduction leads to the disinhibition of neurons in the cerebellar nuclei that receive inhibitory synapses from Purkinje cells (Albus, 1971; Ito et al., 1982; Ito, 2006). The disinhibition 'unlocks' a sequence of neuronal activation steps that eventually enables the contraction of the relevant muscles required for closure of the eyelid.

An impairment of eyeblink conditioning has also been described in ASD mouse models, including mouse models of Fragile X syndrome (Koekkoek et al., 2005), Dup15q syndrome (Piochon et al., 2014; Figure 5.4) as well as mouse models for TSC and Rett syndrome (Kloth et al., 2015). When tested in the same studies, LTD at PF to Purkinje cell synapses was described as abnormal in these mice (Koekkoek et al., 2005; Piochon et al., 2014). It is thus plausible to explain the impairment of classical eyeblink conditioning by LTD deregulation. A causal relationship is particularly supported by the finding in patDp/+ mice, the mouse model for the human 15q11–13 duplication, that CR re-acquisition is normal after successful CR extinction (by repeated application of CS stimuli alone; Figure 5.4) and that LTD induction is normal after prior induction of LTP (Piochon et al., 2014). This finding is in line with the interpretation that bidirectional synaptic plasticity at PF synapses provides an important cellular correlate of some aspects of CR acquisition and extinction (Jörntell and Hansel, 2006).

In contrast to some other motor problems in autism, impairment of eyeblink conditioning is a deficit that does not represent a daily life burden, and it also does not play a role in routine clinical diagnosis of the disorder. The value of this motor learning test lies

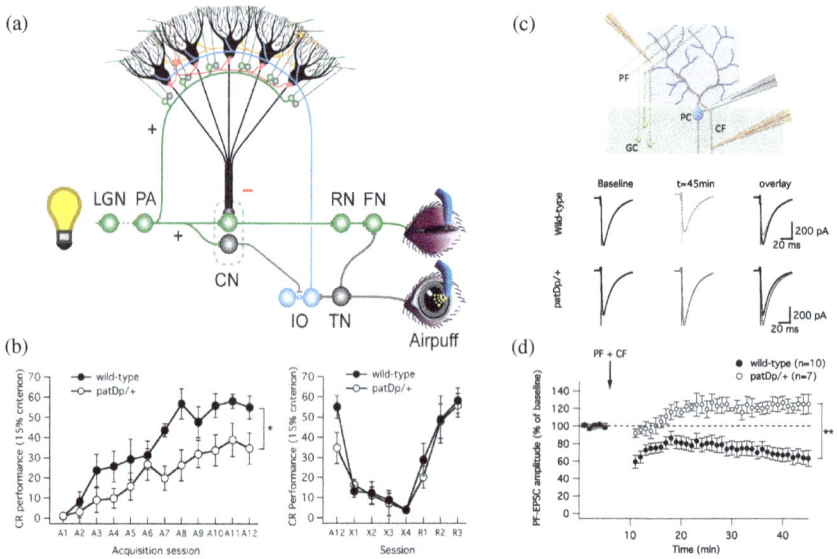

**Figure 5.4:** Classical eyeblink conditioning and LTD are impaired in a mouse model for the human 15q11–13 duplication (patDp/+ mice). (a) Neural circuitry involved in eyeblink conditioning. The US consists of a mild corneal airpuff and the CS of a light signal (or a tone). US signals reach the Purkinje cells (shown in black) via the inferior olive (IO) by CFs, whereas mossy fibers from the pontine area (PA) relay the CS via granule cells (GCs) (small green cells underneath the Purkinje cells) and their axons, the PFs. Repeated presentation of the CS and US results in CRs, during which the eyelid closes in response to the light signal. CN: cerebellar nuclei; FN: facial nucleus; LGN: lateral geniculate nucleus; RN: red nucleus; TN: trigeminal nucleus. (b) Left: Time graph showing CR acquisition (% occurrence) during a sequence of training sessions A1–A12 in wild-type ($n = 11$) and patDp/+ mice ($n = 10$). Note that the 'autistic' mice learn to close their eyes in response to the light signal, but the learning curve shows partial impairment. Right: CR performance during extinction sessions X1–X4, and reacquisition sessions R1–R3. (c) Recording configuration to test for synaptic plasticity in a cerebellar slice preparation. A somatic recording from a Purkinje cell (blue) is performed using a glass pipette (grey). PF burst stimulation is applied by a stimulation electrode (orange, left) placed in the molecular layer. The CF input is activated 120 ms after PF stimulus onset by a second stimulation electrode (orange, right) placed in the GC layer. (d) Top: Typical PF-EPSC traces before and after the application of the tetanization protocol. LTD is induced in wild-type mice, but is absent from patDp/+ mice. Bottom: Time graph showing LTD at PF to Purkinje cell synapses in wild-type mice ($n = 10$) and a potentiation resulting from the same tetanization in patDp/+ mice ($n = 7$). The arrow indicates the time of tetanization. Panel (a) is adapted from Schonewille *et al.* (2010). Panels (b)–(d) are taken from Piochon *et al.* (2014).

elsewhere: first, eyeblink conditioning may in the future serve as an early biomarker for autism, allowing for diagnosis at ages where it is difficult to measure social communication skills (Welsh and Oristaglio, 2016). Second, eyeblink conditioning is conserved throughout mammalian evolution (Fanselow and Poulos, 2005). As I will discuss in more detail in Chapter VI, this is crucial as it allows for a direct comparison of phenotypic consequences in autism between humans and experimental animals. Here, we can conclude that LTD deregulation contributes to a deficit in motor learning. This finding identifies a cellular contributor to motor problems that are typical for autism (proper motor learning is essential for smooth motor performance). But even more notable, LTD deregulation might well represent the first identification of any specific synaptic abnormality that can directly be linked to a behavioral deficit occurring in autism. Notably, cerebellar dysfunction appears to also contribute to social behavior alterations observed in mice (Tsai *et al.*, 2012). This finding supports the view that cerebellar computations do not only orchestrate motor circuits, but might similarly affect cognitive networks as well.

*LTD and the processing of sensory information.* Classical eyeblink conditioning is a prototype of learning the association between two sensory stimuli, such as a light or auditory signal on the one hand, and typically a corneal airpuff on the other. It seems plausible that such associative learning does not only occur in the context of protective motor behaviors, but can be used to form spatial and temporal associations between any two types of sensory input. In cases of cerebellar forms of associative learning, the enormously large number of GC inputs (GCs give rise to PFs and provide 50–80% of all neurons in the brain) and the large number of PF synapses (up to 250,000 synapses per Purkinje cell) provide a neural network that is well equipped for the storage of large numbers of associative 'memories'. In this view, the cerebellum indeed becomes a brain area for sensory processing, and functions in motor control become one of several consequences of cerebellar computation, solely depending on the anatomical organization of cerebellar output structures (Bower, 1997). Several lines of evidence support the view that cerebellar associative learning—and thus underlying plasticity mechanisms including LTD—plays a role in associative learning beyond motor control. First, Purkinje cells respond to a wide range of sensory modalities. In addition to the visual and auditory stimuli used in eyeblink conditioning, they respond to tactile body and whisker stimulation (Thach, 1967; Bosman *et al.*, 2010). And, in electric fish, Purkinje-like cells in the electrosensory lobe (EEL)—a cerebellum-like structure—respond to electric signals (Bell *et al.*, 2008). Thus, the cerebellar cortex receives and processes sensory information from multiple modalities, which can be related in a context-dependent manner in associative learning. Second, the known functions of the cerebellum and cerebellum-like structures in the weakening of predicted sensory input (involving LTD as an underlying mechanism; Bell *et al.*, 1997), as well as in the perception of time intervals (Allman *et al.*, 2014; Teki and Griffith, 2016), provide examples of primarily nonmotor, learning-related computations in the cerebellar cortex.

*Synaptic deregulation and consequences for autism.* LTD deregulation in autism is a phenomenon, whose relevance for ASD phenotypes is not immediately obvious. It is well established that LTD is crucial for synaptic weight regulation and 'synaptic memories' (Hansel and Bear, 2008) processes that—similar to synaptic connectivity changes during development—are essential for circuit plasticity.

Here, I argue that the role of LTD in wiring plasticity and the shaping of brain connectivity explains its relevance in autism (Figure 5.5). LTD deregulation prevents proper associative learning, which is most evident in the impairment of motor learning in autistic patients and ASD mouse models, and might similarly prevent proper processing of sensory inputs.[10] However, the most devastating effect of abnormal LTD pathways is the consequence for developmental synaptic pruning, a process that relies on a largely identical molecular machinery as LTD does. Therefore, pruning will be coimpaired with LTD in the cerebellum, the cerebral cortex, and likely in additional brain areas (Piochon *et al.*, 2016). As illustrated in Figure 5.5, LTD and proper pruning shape and optimize synaptic connectivity.

---

[10]Additional functions of associative learning with relevance to autism—such as in nonmotor aspects of language—will be discussed in Chapter IX.

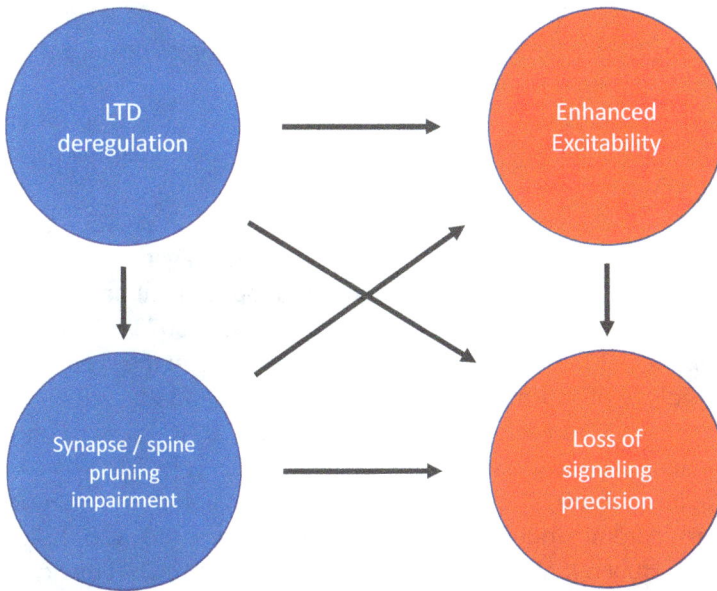

**Figure 5.5:** LTD deregulation and impairment of developmental pruning may lead to further pathological changes, including enhanced excitability and a loss of signaling precision.

Impairment of these functions will lead to higher-order effects, such as enhanced excitability and a loss in the precision of signal processing. LTD deregulation has also been linked to autism-resembling behaviors in mice that I have not yet discussed. Impairment of cerebellar LTD in mutant mice (with a loss of glutamate receptor interaction proteins 1 and 2, Grip1/2, which regulate AMPA receptor trafficking) is associated with increased self-grooming (Mejias *et al.*, 2019). This is a remarkable finding, as this rodent behavior resembles human repetitive behaviors that are described as a core symptom of autism in DSM-5.

# References

Ahmadianteherani, S., and London, S.E. (2017). Bidirectional manipulation of mTOR signaling disrupts socially mediated vocal learning in juvenile songbirds. *Proc. Natl. Acad. Sci. USA* 114, 9463–9468.

Albus, J.S. (1971). A theory of cerebellar function. *Math. Biosci.* 10, 25–61.

Allman, M.J., Teki, S., Griffiths, T.D., and Meck, W.H. (2014). Properties of the internal clock: first- and second-order principles of subjective time. *Annu. Rev. Psychol.* 65, 743–771.

Auerbach, B.D., Osterweil, E.K., and Bear, M.F. (2011). Mutations causing syndromic autism define an axis of synaptic pathophysiology. *Nature* 480, 63–68.

Barnes, S.A., Wijetunge, L.S., Jackson, A.D., Katsanevaki, D., Osterweil, E.K., Komiyama, N.H., Grant, S.G., Bear, M.F., Nägerl, U.V., Kind, P.C., and Wyllie, D.J. (2015). Convergence of hippocampal pathophysiology in Syngap+/– and Fmr1/y mice. *J. Neurosci.* 35, 15073–15081.

Baudouin, S.J. (2014). Heterogeneity and convergence: the synaptic pathophysiology of autism. *Eur. J. Neurosci.* 39, 1107–1113.

Baudouin, S.J., Gaudias, J., Gerharz, S., Hatstatt, L., Zhou, K., Punnakkal, P., Tanaka, K.F., Spooren, W., Hen, R., De Zeeuw, C.I., Vogt, K., and Scheiffele, P. (2012). Shared synaptic pathophysiology in syndromic and non-syndromic rodent models of autism. *Science* 338, 128–132.

Bear, M.F., Huber, K.M., and Warren, S.T. (2010). The mGluR theory of fragile X mental retardation. *Trends Neurosci.* 27, 370–377.

Bell, C.C., Han, V., and Sawtell, N.B. (2008). Cerebellum-like structures and their implications for cerebellar function. *Annu. Rev. Neurosci.* 31, 1–24.

Bell, C.C., Han, V., Sugawara, Y., and Grant, K. (1997). Synaptic plasticity in a cerebellum-like structure depends on temporal order. *Nature* 387, 278–281.

Bosman, L.W., Koekkoek, S.K., Shapiro, J., Rijken, B.F., Zandstra, F., van der Ende, B., Owens, C.B., Potters, J.W., de Gruijl, J.R., Ruigrok, T.J., and De Zeeuw, C.I. (2010). Encoding whisker input by cerebellar Purkinje cells. *J. Physiol.* 588, 3757–3783.

Bower, J.M. (1997). Is the cerebellum sensory for motor's sake, or motor for sensory's sake: the view from whiskers of a rat? *Prog. Brain Res.* 114, 463–496.

Colvert, E., Tick, B., McEwen, F., Stewart, C., Curran, S.R., Woodhouse, E., Gillan, N., Hallett, V., Lietz, S., Garnett, T., Ronald, A., Plomin, R., Rijsdijk, F., Happe, F., and Bolton, P. (2015). Heritability of autism spectrum disorder in a UK population-based twin sample. *JAMA Psychiatry* 72, 415–423.

D'Amato, R.J., Loughnan, M.S., Flynn, E., and Folkman, J. (1994). Thalidomide is an inhibitor of angiogenesis. *Proc. Natl. Acad. Sci. USA* 91, 4082–4085.

Darnell, J.C., Van Driesche, S.J., Zhang, C., Hung, K.Y., Mele, A., Fraser, C.E., Stone, E.F., Chen, C., Fak, J.J., Chi, S.W., Licatalosi, D.D., Richter, J.D., and Darnell, R.B. (2011). FMRP stalls ribosomal translocation on mRNAs linked to synaptic function ad autism. *Cell* 146, 247–261.

Davies, G., *et al.* (2011). Genome-wide association studies establish that human intelligence is highly heritable and polygenic. *Mol. Psychiatry* 16, 996–1005.

Dufour-Rainfray, D., Vourc'h, P., Tourlet, S., Guilloteau, D., Chalon, S., and Andres, C.R. (2011). Fetal exposure to teratogens: evidence of genes involved in autism. *Neurosci. Biobehav. Rev.* 5, 1254–1265.

Fanselow, M.S., and Poulos, A.M. (2005). The neuroscience of mammalian associative learning. *Annu. Rev. Psychol.* 56, 207–234.

Fatemi, S.H., *et al.* (2012). Consensus paper: pathological role of the cerebellum in autism. *Cerebellum* 11, 777–807.

Fingher, N., Dinstein, I., Ben-Shachar, M., Haar, S., Dale, A.M., Eyler, L., Pierce, K., and Courchesne, E. (2017). Toddlers later diagnosed with autism exhibit multiple structural abnormalities in temporal corpus callosum fibers. *Cortex* 97, 291–305.

Fournier, K.A., Hass, C.J., Naik, S.K., Lodha, N., and Cauraugh, J.H. (2010). Motor coordination in autism spectrum disorders: a synthesis and meta-analysis. *J. Autism Dev. Disord.* 40, 1227–1240.

Ganna, A., *et al.* (2019). Large-scale GWAS reveals insights into the genetic architecture of same-sex sexual behavior. *Science* 365, doi: 10.1126/science.aat7693.

Green, D., Charman, T., Pickles, A., Chandler, S., Loucas, T., Simonoff, E., and Baird, G. (2009). Impairment in movement skills of children with autistic spectrum disorders. *Dev. Med. Child Neurol.* 51, 311–316.

Hansel, C. (2019). Deregulation of synaptic plasticity in autism. *Neurosci. Lett.* 688, 58–61.

Hansel, C., and Bear, M.F. (2008). LTD—synaptic depression and memory storage. In Molecular Mechanisms of Memory, in: J.D. Sweatt, J. Byrne (Eds.), Vol. 4 of *Learning and Memory: a Comprehensive Reference*, Elsevier, Oxford, pp. 327–366.

Huber, K.M., Gallagher, S.M., Warren, S.T., and Bear, M.F. (2002). Altered synaptic plasticity in a mouse model for fragile X mental retardation. *Proc. Natl. Acad. Sci. USA* 99, 7746–7750.

Huber, K.M., Klann, E., Costa-Mattioli, M., and Zukin, R.S. (2015). Dysregulation of mammalian target of rapamycin signaling in mouse models of autism. *J. Neurosci.* 35, 13836–13842.

Hutsler, J.J., and Zhang, H. (2010). Increased dendritic spine densities on cortical projection neurons in autism spectrum disorders. *Brain Res.* 1309, 83–94.

Ito, M. (2006). Cerebellar circuitry as a neuronal machine. *Prog. Neurobiol.* 78, 272–303.

Ito, M., Sakurai, M., and Tongroach, P. (1982). Climbing fibre induced depression of both mossy fibre responsiveness and glutamate sensitivity of cerebellar Purkinje cells. *J. Physiol.* 324, 113–134.

Johnson, B.P., Rinehart, N., White, O., Millist, L., and Fielding, J. (2013). Saccade adaptation in autism and Asperger's disorder. *Neuroscience* 243, 76–87.

Jörntell, H., and Hansel, C. (2006). Synaptic memories upside down: bidirectional plasticity at cerebellar parallel fiber—Purkinje cell synapses. *Neuron* 52, 227–238.

Kloth, A.D., Badura, A., Li, A., Cherskov, A., Connolly, S.G., Giovannucci, A., Bangash, M.A., Grasselli, G., Penagarikano, O., Piochon, C., Tsai, P.T., Geschwind, D.H., Hansel, C., Sahin, M., Takumi, T., Worley, P.F., and Wang, S.S. (2015). Cerebellar associative sensory learning defects in five mouse autism models. *eLife* 4, e06085.

Koekkoek, S.K., *et al.* (2005). Deletion of FMR1 in Purkinje cells enhances parallel fiber LTD, enlarges spines, and attenuates cerebellar eyelid conditioning in Fragile X syndrome. *Neuron* 47, 339–352.

Lai, E., *et al.* (2013). An autism-associated neuroligin-3 mutation impairs developmental synapse elimination in the cerebellum. *Soc. Neurosci. Abstract* 718.3.

Landrigan, P.J. (2010). What causes autism? Exploring the environmental contribution. *Curr. Opin. Pediatr.* 22, 219–225.

Markram, H., Rinaldi, K., and Markram, K. (2007). The intense world syndrome—an alternative hypothesis for autism. *Front. Neurosci.* 1, 77–96.

McCormick, D.A., and Thompson, R.F. (1984). Cerebellum: essential involvement in the classically conditioned eyelid response. *Science* 223, 296–299.

Mejias, R., Chiu, S.L., Han, M., Rose, R., Gil-Infante, A., Zhao, Y., Huganir, R.L., and Wang, T. (2019). Purkinje cell-specific *Grip1/2* knockout mice show increased self-grooming and enhanced mGluR5 signaling in cerebellum. *Neurobiol. Dis.* 132, 104602.

Mosconi, M.W., Luna, B., Kay-Stacey, M., Nowinski, C.V., Rubin, L.H., Scudder, C., Minshew, N., and Sweeney, J.A. (2013). Saccade adaptation abnormalities implicate dysfunction of cerebellar-dependent learning mechanisms in Autism Spectrum Disorders (ASD). *PLOS One* 8, e63709.

Mosconi, M.W., Wang, Z., Schmitt, L.M., Tsai, P., and Sweeney, J.A. (2015). The role of cerebellar circuitry alterations in the pathophysiology of autism spectrum disorders. *Front. Neurosci.* 9, 296.

Oguro-Ando, A., Rosensweig, C., Herman, E., Nishimura, Y., Werling, D., Bill, B.R., Berg, J.M., Gao, F., Coppola, G., Abrahams, B.S., and Geschwind, D.H. (2015). Increased CYFIP1 dosage alters cellular and dendritic morphology and dysregulates mTOR. *Mol. Psychiatry* 20, 1069–1078.

Oristaglio, J., Hyman West, S., Ghaffari, M., Lech, M.S., Verma, B.R., Harvey, J.A., Welsh, J.P., and Malone, R.P. (2013). Children with autism spectrum disorders show abnormal conditioned response timing on delay, but not trace, eyeblink conditioning. *Neuroscience* 248, 708–718.

Ornoy, A., Weinstein-Fudim, L., and Ergaz, Z. (2015). Prenatal factors associated with autism spectrum disorder (ASD). *Reprod. Toxicol.* 56, 155–169.

Osterweil, E.K., Krueger, D.D., Reinhold, K., and Bear, M.F. (2010). Hypersensitivity to mGluR5 and ERK1/2 leads to excessive protein synthesis in the hippocampus of a mouse model of Fragile X syndrome. *J. Neurosci.* 30, 15616–15627.

Piochon, C., Kano, M., and Hansel, C. (2016). LTD-like molecular pathways in developmental synaptic pruning. *Nat. Neurosci.* 19, 1299–1310.

Piochon, C., Kloth, A.D., Grasselli, G., Titley, H.K., Nakayama, H., Hashimoto, K., Wan, V., Simmons, D.H., Eissa, T., Nakatani, J., Cherskov, A., Miyazaki, T., Watanabe, M., Takumi, T., Wang, S.S., and Hansel, C. (2014). Cerebellar plasticity and motor learning deficits in a copy-number variation mouse model of autism. *Nat. Commun.* 5, 5586.

Santini, E., Huynh, T.N., MacAskill, A.F., Carter, A.G., Ruggero, D., Pierre, P., Kaphzan, H., and Klann, E. (2013). Exaggerated translation causes synaptic and behavioral aberrations associated with autism. *Nature* 493, 411–415.

Santini, E., and Klann, E. (2014). Reciprocal signaling between translational control pathways and synaptic proteins in autism spectrum disorders. *Sci. Signal.* 7, re10.

Schonewille, M., Belmeguenai, A., Koekkoek, S.K., Houtman, S.H., Boele, H.J., van Beugen, B.J., Gao, Z., Badura, A., Ohtsuki, G., Amerika, W.E., Hosy, E., Hoebeek, F.E., Elgersma, Y., Hansel, C., and De Zeeuw, C.I. (2010). Purkinje cell-specific knockout of the protein phosphatase PP2B impairs potentiation and cerebellar motor learning. *Neuron* 67, 618–628.

Sears, L.L., Finn, P.R., and Steinmetz, J.E. (1994). Abnormal classical eyeblink conditioning in autism. *J. Autism Dev. Disord.* 24, 737–751.

Sestan, N., and State, M.W. (2018). Lost in translation: traversing the complex path from genomics to therapeutics in autism spectrum disorder. *Neuron* 100, 406–423.

Tabuchi, K., Blundell, J., Etherton, M.R., Hammer, R.E., Liu, X., Powell, C.M., and Südhof, T.C. (2007). A neuroligin-3 mutation implicated in autism increases inhibitory synaptic transmission in mice. *Science* 318, 71–76.

Tang, G., *et al.* (2014). Loss of mTOR-dependent macroautophagy causes autistic-like synaptic pruning deficits. *Neuron* 83, 1131–1143.

Teki, S., and Griffiths, T.D. (2016). Brain bases of working memory for time intervals in rhythmic sequences. *Front. Neurosci.* 10, 239.

Thach, W.T. (1967). Somatosensory receptive fields of single units in cat cerebellar cortex. *J. Neurophysiol.* 30, 675–696.

Tsai, P.T., *et al.* (2012). Autistic-like behavior and cerebellar dysfunction in Purkinje cell *Tsc1* mutant mice. *Nature* 488, 647–651.

Wang, M., Li, H., Takumi, T., Qiu, Z., Xu, X., Yu, X., and Bian, W.J. (2017). Distinct defects in spine formation or pruning in two gene duplication mouse models of autism. *Neurosci. Bull.* 33, 143–152.

Wang, S.S., Kloth, A.D., and Badura, A. (2014). The cerebellum, sensitive periods, and autism. *Neuron* 83, 518–532.

Weiler, I.J., and Greenough, W.T. (1993). Metabotropic glutamate receptors trigger postsynaptic protein synthesis. *Proc. Natl Acad. Sci. USA* 90, 7168–7171.

Welsh, J.P., and Oristaglio, J.T. (2016). Autism and classical eyeblink conditioning: performance changes of the conditioned response related to autism spectrum disorder diagnosis. *Front. Psychiatry* 7, 137.

Wolff, J.J., Gerig, G., Lewis, J.D., Soda, T., Styner, M.A., Vachet, C., Botteron, K.N., Elison, J.T., Dager, S.R., Estes, A.M., Hazlett, H.C., Schultz, R.T., Zwaigenbaum, L., Piven, J., and the IBIS Network (2015). Altered corpus callosum morphology associated with autism over the first 2 years of life. *Brain* 138, 2046–2058.

Wood, A.R., *et al.* (2014). Defining the role of common variation in the genomic and biological architecture of adult human height. *Nat. Genetics* 46, 1173–1186.

Zhang, B., Chen, L.Y., Liu, X., Maxeiner, S., Lee, S.J., Gokce, O., and Südhof, T.C. (2015). Neuroligins sculpt cerebellar Purkinje-cell circuits by differential control of distinct classes of synapses. *Neuron* 87, 781–796.

Zhang, M.M., Yu, K., Xiao, C., and Ruan, D.Y. (2003). The influence of developmental periods of sodium valproate exposure on synaptic plasticity in the CA1 region of rat hippocampus. *Neurosci. Lett.* 351, 165–168.

# 6

# The Study of Autism-Resembling Behaviors in Mouse Models

## Modeling Human Behaviors in Laboratory Animals

The use of animals in research on brain developmental disorders, and in particular research on autism, is without alternative, but is also highly problematic. Animal models give access to the study of pathophysiological abnormalities in, for example, synaptic connectivity and plasticity. The electrophysiological and microscopic techniques that are applied in these studies cannot be used in humans. Likewise, mouse models are perfectly suited to examine the consequences of isolated genetic deficits, which are introduced to otherwise normal and healthy mice. The behavioral alterations in these genetically modified mice enable us to establish direct genotype–phenotype interactions. In human subjects, such causalities cannot be as readily established, because any genetic deficit occurs in the context of other, uncontrolled, parameters, such as unknown genetic abnormalities or unknown exposure to environmental factors. In other words, as the genetic manipulation is not introduced experimentally, it also does not necessarily occur in isolation. Unknown factors may contribute to the phenotype observed. Thus, the 'genotype' side of the equation, or the 'cause', is a huge plus when working with genetically modified mice. The other side of the coin, however, the 'phenotype', or the 'consequence', is problematic. This is not about the similarity of synapse physiology or circuit organization between humans and mice. In fact, the structural organization and function of synapses and small circuits are remark-

97

ably similar between mammals. The problem lies in the comparability of behaviors of different mammalian species. The question is not
so much whether wild mice can be autistic or not, as much as the
question is not—to give another example of a brain developmental
disorder—whether mice can be depressed or suicidal. The question
is whether an experimentally introduced genetic manipulation can
cause any behaviors that are similar to those in humans carrying the
same genetic deficit? How can we compare social communication
and interaction, or restricted, repetitive behaviors, between mice and
men? In this chapter,[1] I will introduce some of the mouse behavioral
tests that are used in autism research and explain their strengths and
limitations. Subsequently, I will briefly discuss whether mammalian
species other than mice can provide useful alternatives.

The difficulties in modeling human behaviors in mice begin with
the observation that some human behaviors do not have equivalents
in mice. Autism Spectrum Disorder (ASD) patients may not maintain
normal eye contact during communication and may show aberrant
eye movements (Johnson *et al.*, 2013; Mosconi *et al.*, 2013). Rodents
do not have an equivalent of eye contact. Therefore, it is not possible
to model abnormal eye contact as an ASD phenotype in mice. The
same holds true for verbal communication. However, mice are social
animals, a feature that they share with humans. Therefore, I will
begin by discussing a social interaction test.

## Three-chamber Test for Social Approach

A lack of interest in social interactions is a hallmark characteristic of ASD
(DSM-5). Patients have difficulty reading other people's emotions and do
not enjoy spending time with others. The most commonly used test for
sociality in mice is the three-chamber test, which has several variations
(Figure 6.1). In all configurations, the test mouse is initially placed into
the middle of three serially connected chambers with the end chambers
containing either nothing at all, an object, or a mouse of varied familiarity
(stranger, minimally familiar from one exposure, or familiar).

---

[1]This chapter is in parts adapted from Simmons *et al.* (2020). The author would like
to give credit to Drs. Dana Simmons and Peggy Mason. These colleagues have been
the main contributors to this publication.

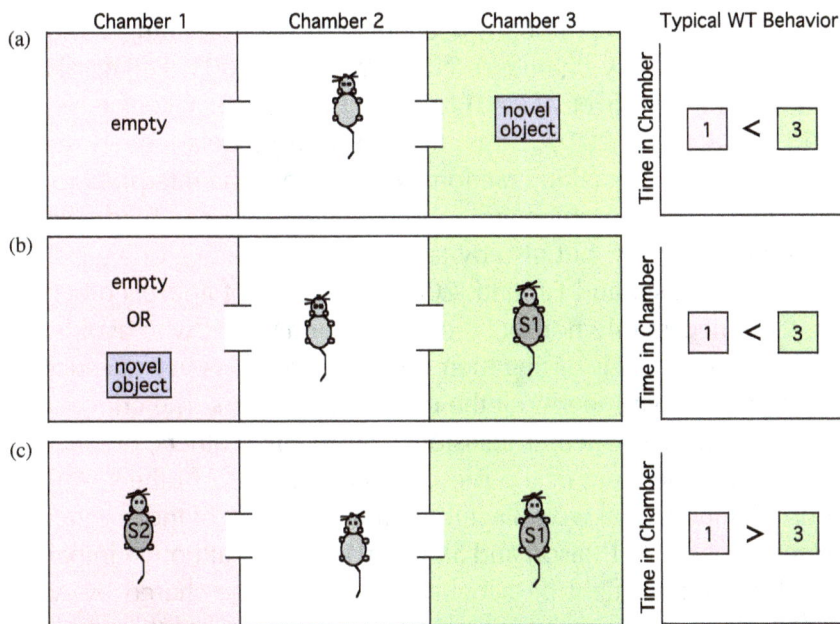

**Figure 6.1:** Variations of the three-chamber test. (a) Control for basic curiosity and exploratory behavior. Left: Experimental configuration with three serially connected chambers. At the beginning of the experiment, the test mouse is placed in the middle chamber (chamber 2). One side chamber is empty and the other contains a novel object. Right: A wild-type mouse will usually spend more time with the object than in the empty chamber. (b) Test for sociality. Left: Experimental configuration with one side chamber either empty or a novel object, and the other containing a stranger mouse (S1). Right: A wild-type mouse typically spends more time with the stranger mouse than with an object or in an empty chamber. (c) Test for novelty effects in social approach behaviors. Left: Experimental configuration with one side chamber containing a new stranger mouse (S2), and the other side chamber containing the familiar stranger mouse (S1). Right: Typical wild-type mice spend more time with the new stranger mouse than with the familiar mouse. This figure is taken from Simmons *et al.* (2020).

After a period of habituation, the test mouse is given access to the side chambers, which are occupied in different constellations, such as 'empty vs. object,' 'stranger vs. object,' or 'stranger vs. familiar.' As illustrated in Figure 6.1, wild-type mice are curious and show a preference to spend time with a novel object over nothing, but also a stranger mouse over an object or a more familiar mouse (e.g. Silverman *et al.*, 2010; Kaidanovich-Beilin *et al.*, 2011). ASD-model mice, in contrast, do not show a preference for a stranger mouse (e.g. over an empty chamber) (Nakatani *et al.*, 2009; Penagarikano *et al.*, 2011; Smith *et al.*, 2011; Tsai *et al.*, 2012; Gkogkas *et al.*, 2013; Sasa *et al.*, 2016). Thus, these mouse models indeed reflect abnormalities in social interaction. Nevertheless, the interpretation of these behaviors is problematic. Approach behavior can be driven by pro-social curiosity, but also by territorial, exploratory, or aggressive motivations (Insel and Fernald, 2004).[2] Absence of approach behavior is not necessarily nonsocial either. An individual who avoids the company of others is basing their behavior on others just as much as the individual who seeks out the company of others. The critical feature of sociality is whether decisions are based on others, or whether they are independent of the presence and actions of others. A truly nonsocial individual would act the same regardless of the absence or presence of others (Mason and Shan, 2017). Thus, although impaired social approach might be a behavioral phenotype shared by ASD-model mice and humans with autism, the strength of the apparent similarity remains unclear. We simply do not know with sufficient confidence the real motivation behind approach behavior in mice. We also do not know whether lack of approach behavior means the same in both humans and mice. Although the three-chamber test for social approach still constitutes a valuable component of a behavioral test suite, the results obtained must be interpreted with caution.

## Ultrasonic Vocalization in Mice

Deficits in communication constitute a diagnostic ASD feature in humans (DSM-5). Notably this is a problem with communication, not necessarily with language. Restricted language, or even the absence of language, is seen in some patients with autism (Sigman and McGovern,

---

[2]Potential aggressive motivations can, to some degree, be addressed by removing barriers between mice at the end of the test period to observe the nature of subsequent interactions (Sankoorial *et al.*, 2006; Fairless *et al.*, 2012) or by allowing mice to engage in free reciprocal interactions in the first place (Silverman *et al.*, 2010).

2005; Baghdadli *et al.*, 2012). Yet, language impairment itself does not constitute one of the diagnostic criteria for ASD. A parallel behavior in mice is ultrasonic vocalization (USV), which models human communication, albeit not language. USVs develop in young mice, which use these sounds to communicate anxiety or fear. This behavior peaks at P5 and is gone by P14 (Noirot, 1966). In adult mice, USVs are used in male–female courtship (Nyby, 2001), but also in territorial disputes between resident and intruder (Moles and D'Amato, 2000). Juveniles from many murine models of ASD show altered USVs in comparison to wild-type mice (15q11-13 duplication, Nakatani *et al.*, 2009; Cnt-nap2$^{-/-}$, Penagarikano *et al.*, 2011; Ube3a overexpressing mice, Smith *et al.*, 2011; L7-Tsc1$^{-/-}$, Tsai *et al.*, 2012; Shank2$^{-/-}$, Won *et al.*, 2012; eIF4Ebp2$^{-/-}$, Gkogkas *et al*, 2013). USV abnormalities have also been characterized in adult ASD-model mice during resident-intruder tests (15q11-13 duplication, Nakatani *et al.*, 2009; Shank2$^{-/-}$, Won *et al.*, 2012) and in male–female courtship (16p11.2 deletion syndrome, Yang *et al.*, 2015).

Similar to the situation that we face with the three-chamber test, there are strengths and limitations when recording USV vocalization in ASD-model mice. An advantage of the inclusion of USVs in the phenotypic characterization of ASD mouse models is that these measures reflect a communication deficit, and that the vocalizations can be readily recorded and analyzed. A caveat, similar to that discussed for social approach, is that the exact function of these USV vocalizations is not known. It is believed that juvenile vocalizations communicate separation stress to the mother (Crawley, 2004). However, there is also evidence for a component of USV distress vocalizations that is related to thermoregulation. USV distress calls are reduced when experiments are kept thermoneutral, that is when separation is not associated with a reduction in body temperature (Blumberg *et al.*, 1992). This importance of thermal consequences of separation suggests that USV differences in juveniles from ASD models may be at least partially related to differences in temperature sensitivity. In adults, it is also difficult to precisely determine the communication functions of USV vocalizations, for example, in different aspects of territorial behavior (Portfors and Perkel, 2014). A useful approach is to combine USV measures with monitoring of scent marking behavior, which defines territorial claims, in the characterization of behavioral phenotypes (Wöhr *et al.*, 2011; Hodges *et al.*, 2017). As a test of phenotypes in ASD-model mice, measures of USV vocalization remain a valuable tool. As we continue to increase our understanding of USV functions in communication, we will get closer to using them to their full scientific potential. For now, though, these tests have to be interpreted with caution.

## Repetitive Behaviors: Marble Burying and Self-grooming

ASD patients show abnormal repetitive and stereotyped behaviors (DSM-5). The latter is a unique feature of autism. Repetitive behaviors partially reflect the high co-morbidity of autism with obsessive-compulsive disorder (OCD), as evident in task-specific perseverations such as repeatedly washing hands or flipping light switches (Soorya *et al.*, 2008; Volkmar, 2009). In mouse models of autism, repetitive behaviors can be observed as well, typically involving scratching, licking, or digging behaviors. Two tests are routinely used for phenotypic characterization: the marble burying test and the self-grooming test.

For the marble burying test, an array of marbles is placed in a cage, and the test mouse is put in the cage for a set period of time (e.g. 20 minutes). During this time, the mouse is allowed to freely interact with the marbles. Mice tend to cover marbles with bedding materials that are available in the cage. Excessive digging and marble burying are evident due to the significantly higher numbers of marbles buried within the test period. This is a straightforward test, and indeed ASD-model mice have been found to bury more marbles than wild-type mice (eIF4Ebp2$^{-/-}$, Gkogkas *et al.*, 2013). Unfortunately, test results are not consistent. There are reports from other mouse models, in which these mice bury the same number of marbles, or fewer marbles, than wild-type mice (16p11.2 deletion and Cntnap2$^{-/-}$, Brunner *et al.*, 2015; L7-Shank2$^{-/-}$, Sasa *et al.*, 2016). It is currently unknown whether this variability in outcome is a true reflection of differences between these genetically altered mice, or rather a reflection of problems with the test itself (e.g. the outcome might be heavily influenced by other parameters, such as anxiety levels or motor problems).

Another commonly used test for repetitive behaviors is the self-grooming test. In this test, the duration of self-cleaning behaviors such as scratching, licking, and nose-wiping are measured within a set period of time (Kalueff et al., 2015). Mice are placed alone in a cage and allowed to habituate to the environment. Then, the time spent on self-grooming is measured, typically for 10 minutes. Studies have consistently found that ASD-model mice spend more time grooming themselves than do wild-type mice (e.g. Cntnap2$^{-/-}$, Penagarikano et al., 2011; Shank3B$^{-/-}$, Peca et al., 2011; Ube3a over-expressing mice; Smith et al., 2011; L7-Tsc1$^{-/-}$, Tsai et al., 2012, and eIF4Ebp2$^{-/-}$, Gkogkas et al., 2013). These results demonstrate that self-grooming is a useful measure of repetitive behaviors in mice. A caveat of this test is that increased self-grooming behavior cannot be exclusively assigned to autism and may also show in OCD.

## Anxiety: Tests for the Occupation of Open Versus Protected Spaces

Anxiety similarly shows a low degree of ASD specificity, as it occurs in many neuropsychiatric disorders, such as in autism, depression, OCD, schizophrenia and, of course, in anxiety disorder itself (Muris et al., 1998). Nevertheless, it is useful to screen for abnormalities in anxiety-driven behaviors, because anxiety is a frequent symptom among ASD patients (van Steensel et al., 2011) and anxiety-reflecting behaviors are common in all mammals. In rodents, two tests—the open field test and the elevated plus maze test—have been used either alone or in combination to quantify anxiety (Hall and Ballachey, 1932; Hall, 1934; Denenberg, 1969; Walf and Frye, 2007).

In the open field test, a mouse is placed in an empty arena and is allowed to freely explore and move around. Test metrics include the time spent in the center versus the time spent in the periphery of the arena. Rodents exhibit thigmotaxis, the tendency to stay close to the edges of a space rather than in the open space in the middle of the arena. While this is a normal protective behavior for rodent species, abnormally high thigmotaxic scores can be interpreted as a result of elevated anxiety. Indeed, ASD-model mice have been shown to spend significantly less time in the center of the arena than wild-type mice (15q11-13 duplication, Tamada *et al.*, 2010; Shank1$^{-/-}$, Silverman *et al.*, 2011).

In the elevated plus maze, a mouse is placed on an elevated platform at the center of a plus sign, formed by the intersection of two open and two closed arms (Walf and Frye, 2007; Komada *et al.*, 2008). Rodents spend most of their time in the closed arms that are bordered by safety railings. Animals that are less anxious venture at a higher rate into the open arms. Multiple ASD mouse models show a significantly higher percentage of time spent in the closed arms than wild-type mice, suggesting increased anxiety (Shank3B$^{-/-}$, Peca *et al.*, 2011; Shank2$^{-/-}$, Won *et al.*, 2012). Yet, other ASD-model mice do not perform differently on the elevated plus maze relative to wild-type mice (Shank1$^{-/-}$, Silverman *et al.*, 2011).

Overall, anxiety tests are a useful addition to any ASD test suite. It needs to be noted, though, that elevated anxiety is not at all specific to autism, and instead is a phenotype shared by several brain developmental and psychiatric disorders.

## Nociception: Hot Plate Test and Tail Flick Test

ASD patients often show hypo- or hypersensitivity to noxious and other stimuli (Allely, 2013; Moore, 2015). Nociception—the perception of painful stimuli—is readily tested in mice. The first test is the hot plate test, in which mice are placed on a hot plate (e.g. heated to 55°C; temperature needs to be low enough not to cause tissue damage) and the latency is measured to either lift their paws or to jump. Importantly, such reaction finishes the experiment, after which the mouse is removed from the plate (Le Bars *et al.*, 2001; Bannon and Malmberg, 2007). This is comparable to a person quickly withdrawing their hand from a hot plate. In ASD-model mice, hyposensitivity (Shank2$^{-/-}$, Ko *et al.*, 2016) or no difference (Shank1$^{-/-}$, Silverman *et al.*, 2011) was reported relative to wild-type mice.

A second test for nociception in mice is the tail flick test (D'Amour and Smith, 1941; Bannon and Malmberg, 2007). A mouse is held so that the last centimeter of its tail is dipped into hot water (e.g. heated to 52°C). The time is measured until the mouse flicks its tail to remove it from the water. As in the hot plate test, the mouse itself stops exposure to the pain-inflicting temperature by its species-typical protective behavior. The validity of this test has been demonstrated by studies that show that flick latency is lengthened by the administration of analgesics such as opioids, and shortened by inflammation or nerve damage (Przewlacki *et al.*, 1999; Wisner *et al.*, 2006). Using the tail flick test, ASD-model mice showed the same outcome as using the hot plate test: Shank2$^{-/-}$ mice showed reduced pain sensitivity (Ko *et al.*, 2016), whereas no effect of genotype was seen in Shank1$^{-/-}$ mice (Silverman *et al.*, 2011).

A major advantage of using the hot plate test or the tail flick test is that they do not rely on assumptions about motivation or feelings, and rather test a simple withdrawal behavior. A caveat is the variability in nociception phenotypes in the human ASD population, the cause of which is not well understood.

## Motor Performance and Motor Learning

The inclusion of tests for motor problems might come as a surprise. After all, autism is primarily known as a cognitive and social disorder; ASD diagnosis, as recommended in DSM-5, is not based on motor impairment. And yet, motor problems are a known component of autism, particularly in children, who often show delays in reaching motor milestones (rolling over, crawling, sitting, standing, walking). About 80% of children with ASD show motor impairment (Green *et al.*, 2009; Fournier *et al.*, 2010). Here are two good reasons to include the motor aspect into mouse studies. First, motor impairment is relatively specific to autism and does not occur in brain developmental or psychiatric disorders with overlapping co-morbidities, such as anxiety disorder, OCD, or depression. This is an important characteristic, because this advantage does not apply to other easily measured behavior types, such as anxiety (see the previous discussion). Second, motor phenotypes can be compared between model animals and humans. As I will discuss in the following, this is particularly true for classical eyeblink conditioning, the motor learning test that I introduced in Chapter V.

Gait is certainly a parameter that differs between two-legged humans and four-legged rodents. Nevertheless, there are gait parameters that are similarly affected in autism. Children with ASD show a tendency for reduced stride length and an increased step width (Kindregan *et al.*, 2015). Both can be considered as adaptations to enhance stability and to compensate for an inability to fine-tune movement. In a study from my own laboratory on gait deviations in a mouse model for the human 15q11-13 duplication (Dup15q syndrome), we observed that these mice have a longer stride length than wild-type mice and an increased step width (Figure 6.2, Piochon *et al.*, 2014). The opposite change in stride length (decreased for humans; increased for mice) might reflect different strategies to increase stability. What is remarkable is that humans and mice use the same change in step width to widen the base of their stance.

In Chapter V, I discussed eyeblink conditioning in detail (Figure 5.4). The power of this motor learning test lies in the fact that this behavior, and its underlying cerebellar substrate, is conserved throughout vertebrate evolution (Fanselow and Poulos, 2005). There are minor differences in kinetic aspects of this type of motor learning, but nevertheless the phenotypic observations can be directly compared between human patients and ASD-model mice. The fact that ASD patients show impairment of eyeblink conditioning (Sears *et al.*, 1994; Oristaglio *et al.*, 2013; Welsh and Oristaglio, 2016), and that eyeblink conditioning is also affected in ASD-model mice (Piochon *et al.*, 2014; Kloth *et al.*, 2015), makes this a powerful test in the phenotypic test suite. Eyeblink conditioning also illustrates the importance of another factor in behavioral experiments that I did not highlight before: the ability to accurately measure and quantify. This adapted behavior presents itself in the successful closure of the eyelid. The motor response can be measured using camera systems or miniature magnet components that are surgically implanted in the upper and lower eyelid. Either way, these devices allow for precise measurements of the smallest eyelid movements and can therefore detect small deviations from normal eyelid closure under these

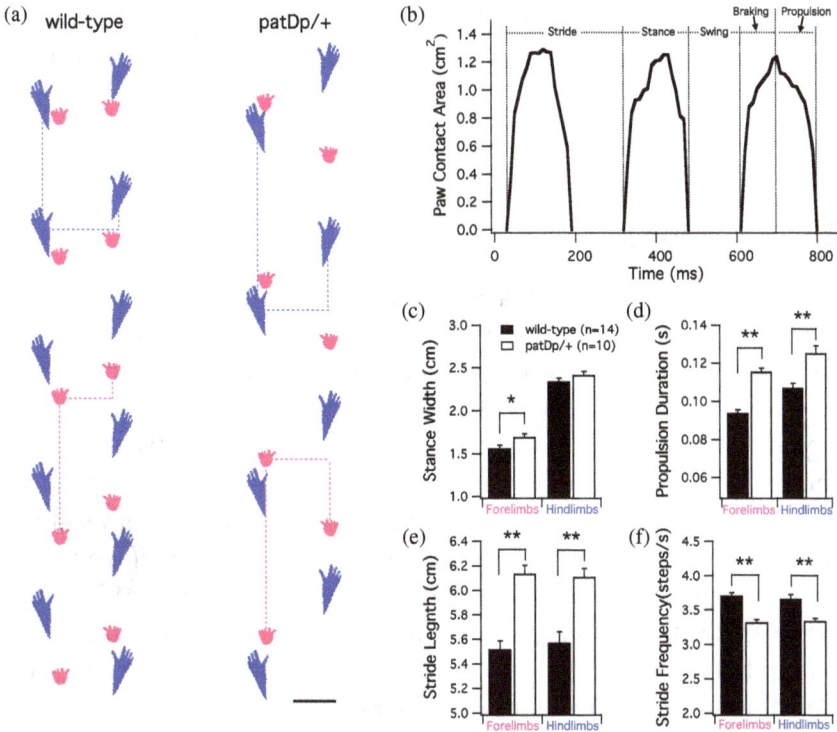

**Figure 6.2:** Gait is impaired in a mouse model for the human 15q11-13 duplication (patDp/+ mice). (a) Example paw prints acquired during locomotion on a treadmill for mice at a running speed of 20 cm s$^{-1}$. Bold dashes indicate the stride length and stance width for both the hindpaws (blue) and forepaws (pink). Scale bar, 1 cm. (b) Example trace of a wild-type mouse paw showing the paw contact area over the course of multiple steps. The stride of a paw can be separated into stance and swing phases. Stance phase (the time the paw is in contact with the ground) is further separated into breaking and propulsion phases, occurring before and after the time of maximal paw contact, respectively. (c–f) Bar graphs showing differences in stance width and stride parameters between patDp/+ ($n = 10$) and wild-type mice ($n = 14$; ** indicates statistical significance of the phenotypic differences). Displayed are the stance width (c), the propulsion duration (d), the stride length (e), and the stride frequency (f). The figure is taken from Piochon $et$ $al.$ (2014).

adapted conditions (in the millisecond range). In comparison, most other tests give relatively crude measures of behavioral phenotypes, for example, time spent with a stranger mouse or time spent in the open center of an arena (in the second range). The simplicity of the motor behavior therefore leads to advantages in comparability, along with precision of measurement and interpretation that are unique to this behavioral test.

## The Importance of Access to Underlying Brain Circuits

In the discussion about behavioral phenotypes in autism, I have so far evaluated a number of different tests that can be applied to mouse models, and this assessment was based on comparability of these behaviors between humans and mice, and on the 'value' of these behaviors with regard to how well they serve as unique markers for autism. There is another crucial factor that needs to be considered. We will not make much progress in our understanding of this complex disorder if we continue to accept that the brain remains a black box and if we ignore the specific changes in brain circuits that ultimately cause ASD symptoms. As illustrated in Figure 5.2, it is significant to understand why certain behavioral alterations result from genetic deficits or environmental influences. Our assessment of behavioral tests, therefore, also needs to consider the question of how well behaviors are understood with regard to underlying circuits and neuronal activity patterns. This is an extremely difficult task, as most behaviors are too complex for such investigation. However, for several ASD-relevant behaviors, we have begun to identify the brain circuits involved, even if often we do not know what aspects of synaptic function and circuit activity are crucial. Social approach behaviors require the co-ordinated activity of several brain centers in the limbic system, such as the ventral tegmental area (VTA) to nucleus accumbens (NAc) projection (Gunaydin et al., 2014), the medial prefrontal cortex (mPFC), hippocampus, and amygdala (Felix-Ortiz and Tye, 2014; Felix-Ortiz et al., 2015; Hung et al., 2017; Murugan et al., 2017). These are brain centers that play a role in emotional control, fear responses, and memory. The experience of social reward depends on oxytocin signaling in the VTA (Hung et al., 2017). A recent publication demonstrates an impairment of this crucial signaling pathway in autism. In dopaminergic neurons in the VTA that lack neuroligin-3, oxytocin signaling is reduced (Hörnberg *et al.*, 2020). In Chapter V, I have introduced a point mutation in neuroligin-3 as one of the known genetic causes of autism (Tabuchi *et al.*, 2007). Therefore, the Hörnberg study provides a rare insight into the neural circuit alterations that may affect social behaviors in ASD.

Repetitive behaviors in mice involve dysfunction of the basal ganglia, which is an important motor center in the brain (Wang *et al.*, 2017). Social vocalizations in mice are controlled by neurons in the periaqueductal gray, which is located in the midbrain (Tschida *et al.*, 2019). These are all relatively recent studies that together illustrate the progress that is made in identifying brain circuits that are critically involved in complex behaviors. And yet, all of these studies share a limitation: it remains unknown how exactly the circuits studied control ASD-relevant behaviors as our understanding of these circuits remains superficial. Classical eyeblink conditioning involves the cerebellum (McCormick and Thompson, 1984). The learning component that leads to eyelid closure upon presentation of the previously neutral conditioned stimulus (CS) depends—at least in part—on LTD at excitatory synapses onto Purkinje cells, a plasticity phenomenon that disinhibits neurons in the cerebellar nuclei and enables motor activation (see Chapter V). Indeed, LTD deregulation is found in ASD-model mice (Koekkoek *et al.*, 2005; Piochon *et al.*, 2014). What is crucial here is the demonstration that one specific cellular defect—dysfunction of a type of synaptic plasticity—can be plausibly linked to an ASD-relevant behavioral phenotype (Figure 6.3). If we are able to similarly identify specific signaling alterations that might be causally related to other behavioral phenotypes, then this would enable us to assess A. whether one or a small group of synaptic (or other) changes takes place that affects different brain areas and

**Figure 6.3:** The study of motor learning deficits allows for a characterization of a chain of events leading to an ASD-typical abnormality. Duplication of chromosome 15q11-13 (Dup15q syndrome) is associated with deregulation of LTD. This synaptic phenotype makes it impossible for the cerebellum to properly adapt its output and ultimately causes motor learning deficits.

impairs behaviors steered by them, or B. whether a large number of diverse cellular alterations leads to the multitude of ASD phenotypes. This knowledge will be crucial for the development of potential treatment strategies.

An example for the relevance of knowledge about synaptic and neuronal abnormalities in autism that does not involve LTD — which I highlighted a lot in this book—has been provided by Mriganka Sur and Rudolf Jaenisch and their research teams at the MIT. Studying Rett syndrome, a brain developmental disorder that is accompanied by autism, Sur and Jaenisch discovered a pharmacological approach for the treatment of some symptoms of the disorder by screening for chemical compounds that lead to an increase in the expression of the $K^+/Cl^-$ cotransporter 2 (KCC2; Tang *et al.*, 2019). Efficient drugs included those that are already approved by the U.S. Food and Drug Administration, and these drugs ameliorated Rett syndrome-associated breathing and locomotion problems in Rett syndrome-model mice (*Mecp2*-mutant mice). This important development has been made possible by a preceding characterization of the cell-physiological consequences of reduced KCC2 expression in cultures of human neurons, namely a deregulation of inhibitory synaptic transmission (mediated by an influx of $Cl^-$ ions into neurons), along with a change in the balance of excitatory and inhibitory transmission (Tang *et al.*, 2016). The transition to a therapeutic approach in humans is still a big challenge. Nevertheless, this promising example demonstrates that true breakthroughs are achieved more easily with detailed knowledge about the pathological changes that a neurodevelopmental disorder causes in the brain itself.

## Animal Models Other than Mice

The vast majority of animal studies on brain developmental and psychiatric disorders are performed using mice. This is because mouse genetic tools have been developed faster than genetic tools for other mammalian species. More recently, genetically modified rats are increasingly available as well, for example, rat models of Fragile X syndrome (Hamilton *et al.*, 2014). The use of rats offers experimental advantages because of their larger size and more diverse behavioral repertoire, which includes complex cognitive behaviors such as empathy and pro-social behaviors (Ben-Ami Bartal *et al.*, 2011) as well as engagement in two-player rat-and-human role-play games (hide-and-seek; Reinhold *et al.*, 2019). In deviation from the murine focus of most biomedical investigations, research groups in China have begun large-scale efforts to use nonhuman primates in ASD research. A study of 10 MeCP2-overexpressing macaque monkeys, mimicking Rett syndrome, revealed increased repetitive behavior, reduced social interactions, and motor deficits as compared to control macaques (Liu *et al.*, 2014, 2016). Shank3-deficient nonhuman primates are also developed (Zhao *et al.*, 2017). The yield of transgenic animals in these studies is low, raising concerns about ethical implications of using large numbers of monkeys. Marmosets, which are small new-world primates that on first impression resemble squirrels more than other primates, are also increasingly pursued for autism research (e.g. Yasue *et al.*, 2018). An advantage of marmosets in ASD research is that they replicate some human behaviors better than other monkeys do. For example, they make nonaggressive eye contact and live in small family structures. Whether or not they share other ASD-relevant behaviors with humans is currently unknown. It is likely that in the near future, these monkey models will be used to address very specific aspects of behavioral alterations in autism that simply cannot be tested in rodents, whereas mouse and rat models will still be used for the majority of studies.

## Autism from a Synaptic Perspective

The central role that synaptopathies—pathological alterations in synaptic function—play in autism and other brain developmental disorders explains the weight given to these conditions in this book. Synaptic plasticity is a phenomenon at the crossroads of developmental brain maturation and adult learning. Therefore, it may not come as a surprise that deregulated plasticity is a recurring theme in autism. The implication of LTD deregulation, in particular in an ASD-typical phenotypic alteration (motor learning) and in synapse/spine pruning deficits, which more generally affect brain connectivity, underscores the catastrophic consequences when these complex regulatory processes are disrupted. In the case of eyeblink conditioning in adult animals, we understand the specific role of LTD quite well and can pinpoint the exact consequences of deregulation. In the case of the involvement of LTD and pruning in connectivity regulation, we have to look at synaptic networks at a larger scale: populations of neurons that are organized in dynamic ensembles ('engrams' in memory functions). In the next chapters, I will discuss the formation and functional relevance of neural ensembles.

# References

Allely, C.S. (2013). Pain sensitivity and observer perception of pain in individuals with autistic spectrum disorder. *Sci. World J.* 2013, 1–20. doi:10.1155/2013/916178.

Baghdadli, A., Assouline, B., Sonie, S., Pernon, E., Darrou, C., Michelon, C., Picot, M.C., Aussilloux, C., and Pry, R. (2012). Developmental trajectories of adaptive behaviors from early childhood to adolescence in a cohort of 152 children with autism spectrum disorders. *J. Autism Dev. Disord.* 42, 1314–1325.

Bannon, A.W., and Malmberg, A.B. (2007). Models of nociception: hot plate, tail-flick, and formalin tests in rodents. *Curr. Protoc. Neurosci.* 41, 8.9.1–8.9.16.

Ben-Ami Bartal, I., Decety, J., and Mason, P. (2011). Empathy and pro-social behavior in rats. *Science* 334, 1427–1430.

Blumberg, M.S., Efimova, I.V., and Alberts, J.R. (1992). Ultrasonic vocalizations by rat pups: the primary importance of ambient temperature and the thermal significance of contact comfort. *Dev. Psychobiol.* 25, 229–250.

Brunner, D., *et al.* (2015). Comprehensive analysis of the 16p11.2 deletion and null Cntnap2 mouse models of autism spectrum disorder. *PLoS One* 8, e0134572.

Crawley, J.N. (2004). Designing mouse behavioral tasks relevant to autistic-like behaviors. *Ment. Retard. Dev. Disabil. Res. Rev.* 10, 248–258.

D'Amour, F.E., and Smith, D.L. (1941). A method for determining loss of pain sensation. *J. Pharmacol. Exp. Ther.* 72, 74–78.

Denenberg, V.H. (1969). Open-field behavior in the rat: what does it mean? *Ann. N.Y. Acad. Sci.* 159, 852–859.

Fairless, A.H., Dow, H.C., Kreibich, A.S., Torre, M., Kuruvilla, M., Gordon, E., Morton, E.A., Tan, J., Berretini, W.H., Li, H., Abel, T., and Brodkin, E.S. (2012). Sociability and brain development in BALB/cJ and C57BL/6J mice. *Behav. Brain Res.* 228, 299–310.

Fanselow, M.S., and Poulos, A.M. (2005). The neuroscience of mammalian associative learning. *Annu. Rev. Psychol.* 56, 207–234.

Felix-Ortiz, A.C., Burgos-Robles, A., Bhagat, N.D., Leppla, C.A., and Tye, K.M. (2015). Bidirectional modulation of anxiety-related and social behaviors by amygdala projections to the medial prefrontal cortex. *Neuroscience* 321, 197–209.

Felix-Ortiz, A.C., and Tye, K.M. (2014). Amygdala inputs to the ventral hippocampus bidirectionally modulate social behavior. *J. Neurosci.* 34, 586–595.

Fournier, K.A., Hass, C.J., Naik, S.K., Lodha, N., and Cauraugh, J.H. (2010). Motor coordination in autism spectrum disorders: a synthesis and meta-analysis. *J. Autism Dev. Disord.* 40, 1227–1240.

Gkogkas, C.G., Khoutorsky, A., Ran, I., Rampakakis, E., Nevarko, T., Weatherill, D.B., Vasuta, C., Yee, S., Truitt, M., Dallaire, P., Major, F., Lasko, P., Ruggero, D., Nader, K., Lacaille, J.C., and Sonenberg, N. (2013). Autism-related deficits via dysregulated eIF4E-dependent translational control. *Nature* 493, 371–377.

Green, D., Charman, T., Pickles, A., Chandler, S., Loucas, T., Simonoff, E., and Baird, G. (2009). Impairment in movement skills of children with autistic spectrum disorders. *Dev. Med. Child Neurol.* 51, 311–316.

Gunaydin, L.A., Grosenick, L., Finkelstein, J.C., Kauvar, I.V., Fenno, L.E., Adhikari, A., Lammel, S., Mirzabekov, J.J., Airan, R.D., Zalocusky, K.A., Tye, K.M., Anikeeva, P., Malenka, R.C., and Deisseroth, K. (2014). Natural neural projection dynamics underlying social behavior. *Cell* 157, 1535–1551.

Hall, C.S. (1934). Emotional behavior in the rat. I. Defecation and urination as measures of individual differences in emotionality. *J. Comp. Psychol.* 18, 385–403.

Hall, C.S., and Ballachey, E.L. (1932). *A Study of the Rat's Behavior in a Field: a Contribution to Method in Comparative Psychology* (University of California Publications in Psychology), vol. 6, pp. 1–12.

Hamilton, S.M., Green, J.R., Veeraragavan, S., Yuva, L., McCoy, A., Wu, Y., Warren, J., Little, L., Ji, D., Cui, X., Weinstein, E., and Paylor, R. (2014). Fmr1 and Nlgn3 knockout rats: novel tools for investigating autism spectrum disorders. *Behav. Neurosci.* 128, 103–109.

Hodges, S.L., Nolan, S.O., Reynolds, C.D., and Lugo, J.N. (2017). Spectral and temporal properties of calls reveal deficits in ultrasonic vocalizations of adult Fmr1 knockout mice. *Behav. Brain Res.* 332, 50–58.

Hörnberg, H., Perez-Garci, E., Schreiner, D., Hatstatt-Burkle, L., Magara, F., Baudouin, S., Matter, A., Nacro, K., Pecho-Vrieseling, E., and Scheiffele, P. (2020). Rescue of oxytocin response and social behavior in a mouse model of autism. *Nature*, 584, 252–256.

Hung, L.W., Neuner, S., Polepalli, J.S., Beier, K.T., Wright, M., Walsh, J.J., Lewis, E.M., Luo, L., Deisseroth, K., Dölen, G., and Malenka, R.C. (2017). Gating of social reward by oxytocin in the ventral tegmental area. *Science* 357, 1406–1411.

Insel, T.R., and Fernald, R.D. (2004). How the brain processes social information: searching for the social brain. *Annu. Rev. Neurosci.* 27, 697–722.

Johnson, B.P., Rinehart, N., White, O., Millist, L., and Fielding, J. (2013). Saccade adaptation in autism and Asperger's disorder. *Neuroscience* 243, 76–87.

Kaidanovich-Beilin, O., Lipina, T., Vukobradovic, I., Roder, J., and Woodgett, J.R. (2011). Assessment of social interaction behaviors. *J. Vis. Exp.* 48, e2473.

Kalueff, A.V., Stewart, A.M., Song, C., Berridge, K.C., Graybiel, A.M., and Fentress, J.C. (2015). Neurobiology of rodent self-grooming and its value for translational neuroscience. *Nat. Rev. Neurosci.* 17, 45–59.

Kindregan, D., Gallagher, L., and Gormley, J. (2015). Gait deviations in children with autism spectrum disorders: a review. *Autism Res. Treat.* 2015, e741480.

Kloth, A.D., Badura, A., Li, A., Cherskov, A., Connolly, S.G., Giovannucci, A., Bangash, M.A., Grasselli, G., Penagarikano, O., Piochon, C., Tsai, P.T., Geschwind, D.H., Hansel, C., Sahin, M., Takumi, T., Worley, P.F., and Wang, S.S. (2015). Cerebellar associative sensory learning defects in five mouse autism models. *eLife* 4, e06085.

Ko, H.G., Oh, S.B., Zhuo, M., and Kaang, B.K. (2016). Reduced acute nociception and chronic pain in Shank2-/- mice. *Mol. Pain* 12, 1–5.

Komada, M., Takao, K., and Miyakawa, T. (2008). Elevated plus maze for mice. *J. Vis. Exp.* 22, e1088.

Le Bars, D., Gozariu, M., and Cadden, S.W. (2001). Animal models of nociception. *Pharmacol. Rev.* 53, 597–652.

Liu, H., *et al.* (2014). TALEN-mediated gene mutagenesis in rhesus and cynomolgus monkeys. *Cell Stem Cell.* 14.3, 323–328.

Liu, Z., *et al.* (2016). Autism-like behaviours and germline transmission in transgenic monkeys overexpressing MeCP2. *Nature.* 530, 98–102.

Mason, P., and Shan, H. (2017). A valence-free definition of sociality as any violation of inter-individual independence. *Proc. Biol. Sci.* 284, 1866.

McCormick, D.A., and Thompson, R.F. (1984). Cerebellum: essential involvement in the classically conditioned eyelid response. *Science* 223, 296–299.

Moles, A., and D'Amato, F.R. (2000). Ultrasonic vocalization by female mice in the presence of a conspecific carrying food cues. *Anim. Behav.* 60, 689–694.

Moore, D.J. (2015). Acute pain experience in individuals with autism spectrum disorders: a review. *Autism* 19, 387–399.

Mosconi, M.W., Luna, B., Kay-Stacey, M., Nowinski, C.V., Rubin, L.H., Scudder, C., Minshew, N., and Sweeney, J.A. (2013). Saccade adaptation abnormalities implicate dysfunction of cerebellar-dependent learning mechanisms in Autism Spectrum Disorders (ASD). *PLoS One* 8, e63709.

Muris, P., Steerneman, P., Merckelbach, H., Holdrinet, I., and Meesters, C. (1998). Comorbid anxiety symptoms in children with pervasive developmental disorders. *J. Anxiety Disord.* 12, 387–393.

Murugan, M., Jang, H.J., Park, M., Miller, E., Taliaferro, J., Cox, J., Parker, N.F., Bhave, V., Nectow, A., Pillow, J.W., and Witten, I.B. (2017). Combined social and spatial coding in a descending projection from the prefrontal cortex. *Cell* 171, 1663–1677.

Nakatani, J., *et al.* (2009). Abnormal behavior in a chromosome-engineered mouse model for human 15q11-13 duplication seen in autism. *Cell* 137, 1235–1246.

Noirot, E. (1966). Ultra-sounds in young rodents. I. Changes with age in albino mice. *Anim. Behav.* 14, 459–462.

Nyby, J.G. (2001). Auditory communication among adults. In: Willott, J.F. (Ed.), *Handbook of Mouse Auditory Research: From Behavior to Molecular Biology* (New York: CRC), pp. 3–18.

Oristaglio, J., Hyman West, S., Ghaffari, M., Lech, M.S., Verma, B.R., Harvey, J.A., Welsh, J.P., and Malone, R.P. (2013). Children with autism spectrum disorders show abnormal conditioned response timing on delay, but not trace, eyeblink conditioning. *Neuroscience* 248, 708–718.

Peca, J., Feliciano, C., Ting, J.T., Wang, W., Wells, M.F., Venkatraman, T.N., Lascola, C.D., Fu, Z., and Feng, G. (2011). Shank3 mutant mice display autistic-like behaviors and striatal dysfunction. *Nature* 472, 437–442.

Penagarikano, O., Abrahams, B.S., Herman, E.I., Winden, K.D., Gdalyahu, A., Dong, H., Sonnenblick, L.I., Gruver, R., Almajano, J., Bragin, A., Golshani, P., Trachtenberg, J.T., Peles, E., and Geschwind, D.H. (2011). Absence of CNTNAP2 leads to epilepsy, neuronal migration abnormalities, and core autism-related deficits. Cell 147, 235–246.

Piochon, C., Kloth, A.D., Grasselli, G., Titley, H.K., Nakayama, H., Hashimoto, K., Wan, V., Simmons, D.H., Eissa, T., Nakatani, J., Cherskov, A., Miyazaki, T., Watanabe, M., Takumi, T., Wang, S.S., and Hansel, C. (2014). Cerebellar plasticity and motor learning deficits in a copy-number variation mouse model of autism. Nat. Commun. 5, 5586.

Portfors, C.V., and Perkel, D.J. (2014). The role of ultrasonic vocalizations in mouse communications. Curr. Opin. Neurobiol. 28, 115–120.

Przewlacki, R., Labuz, D., Mika, J., Przewlocka, B., Tomboly, C., and Toth, G. (1999). Pain inhibition by endomorphins. Ann. N.Y. Acad. Sci. 897, 154–164.

Reinhold, A.S., Sanguinetti-Scheck, J.I., Hartmann, K., and Brecht, M. (2019). Behavioral and neural correlates of hide-and-seek in rats. Science 365, 1180–1183.

Sankoorial, G.M., Kaercher, K.A., Boon, C.J., Lee, J.K., and Brodkin, E.S. (2006). A mouse model system for genetic analysis of sociability: C57BL/6J versus BALB/cJ inbred mouse strains. Biol. Psychiatry 59, 415–423.

Sasa, P., ten Brinke, M.M., Stedehouder, J., Reinelt, C.M., Wu, B., Zhou, H., Zhou, K., Boele, H.J., Kushner, S.A., Lee, M.G., Schmeisser, M.J., Boeckers, T.M., Schonewille, M., Hoebeek, F.E., and De Zeeuw, C.I. (2016). Dysfunctional cerebellar Purkinje cells contribute to autism-like behavior in Shank2-deficient mice. Nat. Commun. 7, 12627.

Sears, L.L., Finn, P.R., and Steinmetz, J.E. (1994). Abnormal classical eyeblink conditioning in autism. J. Autism Dev. Disord. 24, 737–751.

Sigman, M., and McGovern, C.W. (2005). Improvement in cognitive and language skills from preschool to adolescence in autism. J. Autism and Dev. Disord. 35, 15–23.

Silverman, J.L., Yang, M., Lord, C., and Crawley, J.N. (2010). Behavioral phenotyping assays for mouse models of autism. Nat. Rev. Neurosci. 11, 490–502.

Simmons, D.H., Titley, H.K., Hansel, C., and Mason, P. (2020). Behavioral tests for mouse models of autism: an argument for the inclusion of cerebellum-controlled motor behaviors. Neuroscience, doi: 10.1016/j.neuroscience.2020.05.010

Smith, S.E., Zhou, Y.D., Zhang, G., Jin, Z., Stoppel, D.C., and Anderson, M.P. (2011). Increased gene dosage of *Ube3a* results in autism traits and decreased glutamate synaptic transmission in mice. *Sci. Transl. Med.* 3, 103.

Soorya, L., Kiarashi, J., and Hollander, E. (2008). Psychopharmacologic interventions for repetitive behaviors in autism spectrum disorders. *Child Adolesc. Psychiatr. Clin. N. Am.* 17, 753–771.

Tabuchi, K., Blundell, J., Etherton, M.R., Hammer, R.E., Liu, X., Powell, C.M., and Südhof, T.C. (2007). A neuroligin-3 mutation implicated in autism increases inhibitory synaptic transmission in mice. *Science* 318, 71–76.

Tamada, K., Tomonaga, S., Hatanaka, F., Nakai, N., Takao, K., Miyakawa, T., Nakatani, J., and Takumi, T. (2010). Decreased exploratory activity in a mouse model of 15q duplication; implications for disturbance of serotonin signaling. *PLoS One* 5, e15126.

Tang, X., Drotar, J., Li, K., Clairmont, C.D., Brumm, A.S., Sullins, A.J., Wu, H., Liu, X.S., Wang, J., Gray, N.S., Sur, M., and Jaenisch, R. (2019). Pharmacological enhancement of *KCC2* gene expression exerts therapeutic effects on human Rett syndrome neurons, and *Mecp2* mutant mice. *Sci. Transl. Med.* 11, eaau0164.

Tang, X., Kim, J., Zhou, L., Wengert, E., Zhang, L., Wu, Z., Carromeu, C., Muotri, A.R., Marchetto, M.C., Gage, F.H., and Chen, G. (2016). KCC2 rescues functional deficits in human neurons derived from patients with Rett syndrome. *Proc. Natl. Acad. Sci. USA* 113, 751–756.

Tsai, P.T., Hull, C., Chu, Y., Greene-Colozzi, E., Sadowski, A.R., Leech, J.M., Steinberg, J., Crawley, J.N., Regehr, W.G., and Sahin, M. (2012). Autistic-like behavior and cerebellar dysfunction in Purkinje cell Tsc1 mutant mice. *Nature* 488, 647–651.

Tschida, K., Michael, V., Takatoh, J., Han, B.X., Zhao, S., Sakurai, K., Mooney, R., and Wang, F. (2019). A specialized neural circuit gates social vocalizations in the mouse. *Neuron* 103, 459–472.

Van Steensel, F.J., Boegels, S.M., and Perrin, S. (2011). Anxiety disorders in children and adolescents with autistic spectrum disorders: a meta-analysis. *Clin. Child Fam. Psychol. Rev.* 14, 302–317.

Volkmar, F.R. (2009). Citalopram treatment in children with autism spectrum disorders and high levels of repetitive behavior. *Arch. Gen. Psychiatry* 66, 581–582.

Walf, A.A., and Frye, C. (2007). The use of the elevated plus maze as an assay of anxiety-related behavior in rodents. *Nat. Protoc.* 2, 322–328.

Wang, W., et al. (2017). Striatopallidal dysfunction underlies repetitive behavior in Shank3-deficient model of autism. *J. Clin. Invest.* 127, 1978–1990.

Welsh, J.P., and Oristaglio, J.T. (2016). Autism and classical eyeblink conditioning: performance changes of the conditioned response related to autism spectrum disorder diagnosis. *Front. Psychiatry* 7, 137.

Wisner, A., Dufour, E., Messaoudi, M., Nejdi, A., Marcel, A., Ungeheuer, M.N., and Rougeot, C. (2006). Human opiorphin, a natural antinociceptive modulator of opioid-dependent pathways. *Proc. Natl. Acad. Sci. USA* 103, 17979–17984.

Wöhr, M., Roullet, F.I., Hung, A.Y., Sheng, M., and Crawley, J.N. (2011). Communication impairments in mice lacking Shank1: reduced levels of ultrasonic vocalizations and scent marking behavior. *PLoS One* 6, e20631.

Won, H., et al. (2012). Autistic-like social behavior in Shank2-mutant mice improved by restoring NMDA receptor function. *Nature* 486, 261–265.

Yang, M., Mahrt, E.J., Lewis, F., Foley, G., Portmann, T., Dolmetsch, R.E., Portfors, C.V., and Crawley, J.N. (2015). 16p11.2 deletion syndrome mice display sensory and ultrasonic vocalization deficits during social interactions. *Autism Res.* 8, 507–521.

Yasue, M., Nakagami, A., Nakagaki, K., Ichinohe, N., and Kawai, N. (2018). Inequity aversion is observed in common marmosets but not in marmoset models of autism induced by prenatal exposure to valproic acid. *Behav. Brain Res.* 343, 36–40.

Zhao, H., Tu, Z., Xu, H., Yan, S., Yan, H., Zheng, Y., Yang, W., Zheng, J., Li, Z., Tian, R., Lu, Y., Guo, X., Jiang, Y.H., Li, X.J., and Zhang, Y.Q. (2017). Altered neurogenesis and disrupted expression of synaptic proteins in prefrontal cortex of SHANK3-deficient non-human primate. *Cell Res.* 27, 1293–1297.

# 7  The Memory Engram

The role of synapses in the communication between neurons along with their ability to undergo activity-dependent plasticity are factors that identify synaptic contacts as crucial players in learning. The representation of learned information ultimately requires the activity of neuronal ensembles. The term 'ensemble' characterizes a population of neurons that work together toward a shared purpose. In the more specific case of memory storage, ensembles become 'engrams.' A memory engram (see Figure 0.1) encodes information, which is 'represented' by the synchronized activity of this group of neurons. The term 'engram' and its conceptual framework were introduced in 1904 by the German zoologist Richard Semon (Figure 7.1) in his book 'Die Mneme' (The Mneme).[1] I first cite his definition in the original German, and offer my translation into English as follows:

> 'In sehr vielen Fällen lässt sich nachweisen, dass die reizbare Substanz des Organismus, gehöre er nun dem Protisten-, Pflanzen- oder Tierreich an, nach Einwirkung und Wiederaufhören eines Reizes und nach Wiedereintritt in den sekundären Indifferenzzustand dauernd verändert ist. Ich bezeichne diese Wirkung der Reize als ihre engraphische Wirkung, weil sie sich in die organische Substanz sozusagen eingräbt oder einschreibt. Die so bewirkte Veränderung der organischen Substanz bezeichne ich als das Engramm des betreffenden Reizes, und die Summe der Engramme, die ein Organismus besitzt, als seinen Engrammschatz, wobei ein ererbter von einem individuell erworbenen Engrammschatz zu unterscheiden

---

[1] Semon, R. (1904). *Die Mneme als erhaltenes Prinzip im Wechsel des organischen Geschehens* (Leipzig: Verlag von Wilhelm Engelmann).

*ist. [....] Den Inbegriff der mnemischen Fähigkeiten eines Organis-*
*mus bezeichne ich als seine Mneme'*[3]

*English translation*

'In many cases, it can be demonstrated that the excitable substance of the organism, be it a protist,[4] a plant or an animal, after
presentation and removal of a stimulus, and after reentry into a
secondary condition of indifference, is permanently changed. I
describe this effect of stimuli as their engraphic effect, because
it is engraved or inscripted into the organic substance. I call
the resulting change in organic substance the engram of the
respective stimulus, and the sum of all engrams that an organism pos-
sesses its engram portfolio, in which an inherited engram portfolio
needs to be distinguished from an individually acquired one. [....] I
call the epitome of mnemic capabilities of an organism its mneme'.*

Semon was the first to talk about a physical substrate of memory,
a memory trace. He saw the need to introduce a new terminology
that differed from terms describing memory as a cognitive process
(e.g. 'remember'). It is likely that he emphasized the biological substrate of memory (see our discussion of 'circuit learning' in the
Introduction), because he expanded the concept to genetics, claiming that inherited or acquired genetic traits would also form engrams,
very much like stimuli could engrave memory traces in the brain.
Although the hypothesis of the heritability of acquired traits has not
withstood the test of time, Semon's treatment of organismal history—
phylogenic or ontogenic in nature—as an 'engram' introduced an
entirely new perspective on learning. To this day, it provides the theoretical foundation for any substrate- rather than behavior-based
approach to the phenomenon of learning. Semon was an

---

[2]Schacter, D.L., Eich, J.E., and Tulving, E. (1978). Richard Semon's theory of
memory. *J. V. L. V. B.* 17, 721–743. Goeschel, C. (2009). *Suicide in Nazi
Germany* (Oxford University Press).
[3]Cited from the second edition of the book (1908), in which Semon slightly
modified his definition of the Mneme. The term 'Mneme' itself originates
from the Greek mythology, were Mneme was the muse of memory.
[4]A protist is a eukaryotic organism that is not an animal, plant, or fungus.

**Figure 7.1:**   Richard Semon (1859–1918). Semon was born into a Jewish family in Berlin. A zoologist by training, he became an evolutionary biologist at the University of Jena, but later gave up his professorship at Jena as a consequence of a love affair and moved, as a private scholar, to Munich. Throughout his professional career, Semon suffered from the lack of interest in his work by contemporary scholars. In December 1918, he committed suicide, wrapped in a German Imperial flag, devastated by the German defeat in World War I that happened a few weeks earlier, and the death of his wife at Easter the same year.[2]

unconventional and brilliant scholar. Today, thanks to our renewed interest in engrams, his impact on the field finally receives much-deserved recognition.

Since the discovery of LTP by Tim Bliss and Terje Lomo in 1973 (Bliss and Lomo, 1973), synaptic plasticity was widely accepted as a central mechanism in engram formation. Memory research has for many years subsequent to this finding focused on the synapse as the ultimate location of learning-related changes, rather than on the question what the nature and location of the more extensive engram may be—including the many synapses and neurons that likely make up a complete engram. This was perhaps a consequence of the 'failed' engram search experiments conducted by the American psy-chologist Karl Lashley in the 1920s, who systematically lesioned distinct areas of the cortex after training rats to navigate in a maze. Lashley observed that these cortical lesions indeed impaired maze learning, but there was no effect of the location of the lesion (summarized in Lashley, 1950). These findings allowed for the conclusion that engrams are—at least to some degree—distributed, and not confined to a restricted location. As I will describe in the following, new recording and manipulation techniques have recently become available that enabled a renaissance of engram physiology. In the 50 years and more in between, however, the engram remained an elusive concept.

Synapses, in contrast, became technically accessible, and properties of their plasticity indicated a critical involvement in learning. These properties include the activity- and experience-dependence of synaptic plasticity, as well as its ability to drive spike firing in a given population of neurons. Synaptic plasticity has been largely studied ex vivo, and with a focus on the cellular machinery that enables potentiation. This changed in the early 2000s with the arrival of new experimental recording and manipulation techniques, such as optogenetics, that provided access to the study of plasticity and engram physiology in living research animals. In 2014, Roberto Malinow and his colleagues at the University of California at San Diego (UCSD) used these techniques to explore a causal relationship between synaptic plasticity and memory formation (Nabavi *et al.*, 2014). Malinow and colleagues used fear-conditioning, a fast and robust type of learning that involves the amygdala, as a model for behavioral learning. In the classical configuration of this learning test in rodents, a neutral conditioned stimulus (CS), often a tone, is paired with a mild electric shock as the unconditioned stimulus (US). The learned behavior is manifested as freezing of the animal. During these particular experiments, freezing was quantified as an interruption in another type of motor behavior, the trained and rewarded pressing of a lever. In a next step, the researchers replaced the tone presentation with optical stimulation, via an implanted cannula, of neural inputs to the amygdala that originated in an auditory relay nucleus, thus stimulating the same type of axons that would be activated by playing a tone. This optogenetic activation was made possible by injecting a virus into the auditory nucleus that expressed channelrhodopsin 2 (ChR2). The advantage to direct stimulation of axon bundles by light is that it can be controlled which synapses get activated. Application of a tone, in contrast, may lead to activity patterns and uncontrolled plasticity effects throughout the brain. To test whether synaptic plasticity alone could control fear memories, Malinow and colleagues first initiated acquisition of a conditioned response (CR) by pairing a light pulse with the US stimulus. Next, they applied LTD-inducing light stimuli to these axons and observed the inactivation of the previously established memory. Subsequent application of LTP-inducing stimuli reactivated the appropriate fear memory (Figure 7.2). The ability of these specific light patterns to elicit LTD and LTP, respectively, was tested in separate electrophysiological recordings in vivo. These experiments are now considered the most direct evidence for a causal role of synaptic plasticity in memory formation (hence the title of the paper, 'Engineering a memory with LTD and LTP'). A later study has confirmed and expanded on these findings, showing synapse and memory specificity: memories and their recall probability could be modulated by optogenetic potentiation or depotentiation of synapses assigned to a specific memory (Abdou *et al.*, 2018).

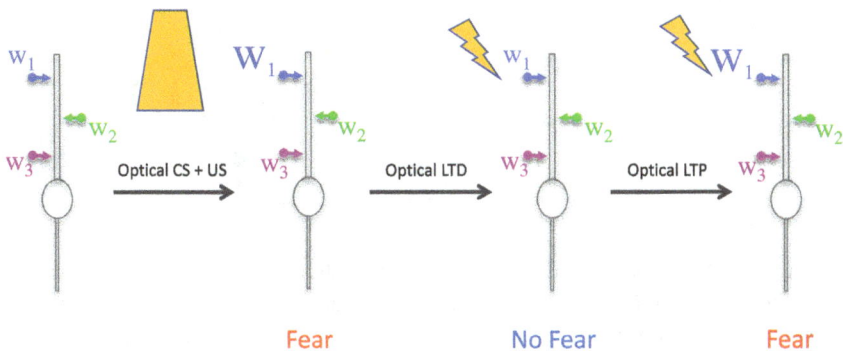

**Figure 7.2:** Illustration of the experiment on the optical control of memories by Malinow and colleagues. Synaptic weights are illustrated by a 'W' and its respective letter size. The optical CS is symbolized by a wide light 'beam' (left), whereas the more specific optical LTD and LTP stimuli are symbolized by light flashes (middle and right).

Optogenetic techniques have also been successfully used to identify the neuronal populations that make up an engram by labeling those neurons that were active during the encoding of a memory, and then testing whether reactivation of the very same neurons elicits the learned behavioral response. I will describe here a strategy to achieve this goal that is used by several laboratories, in particular that of Susumu Tonegawa at MIT. The approach is illustrated in Figure 7.3. There are two components: labeling of neurons that are active in the context of a learning task and reactivation. Labeling is achieved by the (functional) coupling of the genetic promoter of *c-fos* to the expression of a fluorescent protein, EYFP ('enhanced yellow fluorescent protein,' a variant of the classical GFP, 'green fluorescent protein'). *c-fos* is an immediate early gene, whose expression levels increase upon a variety of cellular stimuli, including neuronal activity. Thus, using the promoter region of *c-fos*, but instead coupling it

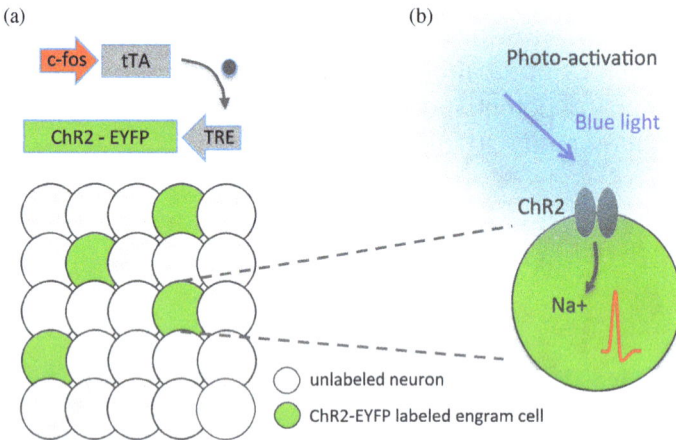

**Figure 7.3:** Memory engram labeling and reactivation. (a) Engram cells are labeled by expression of ChR2-EYFP after activation of an activity-dependent promoter such as *c-fos*. Labeled engram cells (green) can be reactivated by photo-stimulation of channelrhodopsin (ChR2). TRE, tetracycline-responsive element; tTA, tetracycline transactivator (gray dot). (b) Photoactivation of a ChR2-expressing neuron with blue light causes influx of Na+ ions and depolarization, which brings the membrane potential closer to the spike threshold and enhances the probability of spike firing. This figure is adapted from Titley *et al.*, 2017.

to a fluorescent marker, will enhance fluorescence in activated neurons. If now the protein whose expression is enhanced is not only a fluorescent protein, but a fusion protein between EYFP and ChR2, the opsin that I introduced before as the light-stimulated activator of neurons in optogenetic applications, the activity-dependent labeling of engram neurons (via EYFP expression) and the ability to activate neurons by light (via ChR2 expression), are achieved in one step of molecular manipulation: the functional linkage of the fusion protein ChR2-EYFP to the *c-fos* promoter. There is one more built-in feature. To make sure that the activity-dependent expression of ChR2-EYFP is limited to the specific learning period, rather than the entire lifespan of the mouse before the learning experiment begins, the expression of the fusion protein is made conditional to the absence of a chemical substance, doxycycline (Dox), which can simply be added to or removed from the diet. In this gene induction system (Reijmers *et al.*, 2007), ChR2-EYFP is not physically coupled to c-fos. Instead, it is activated by the diffusible product of another gene called tetracycline transactivator (tTA), which is now coupled to the *c-fos* promoter. Its protein product, tTA, binds to its target tetracycline-responsive element (TRE promoter), but this binding is prevented in the presence of Dox. Tonegawa and colleagues targeted the dentate gyrus (a subregion of the hippocampus) of c-fos-tTA mice with a virus containing a TRE-ChR2-EYFP construct. Under these conditions, it is possible to limit the increase of ChR2-EYFP expression in response to c-fos activation to periods when Dox is removed from the diet of the mouse. If this is timed to coincide with the learning phase, only those neurons expressing ChR2-EYFP were activated during the learning task.

Susumu Tonegawa and his colleagues used this technique to identify the memory engram in hippocampal dentate gyrus when mice were trained in a fear-conditioning paradigm (Liu *et al.*, 2012). In this experiment, mice were first habituated to a specific environment (here called 'context') A. During the time of habituation (5 days), Dox was added to the diet. Subsequently, the mice were taken off Dox for 2 days, and fear-conditioned on day 3. During these 3 days, the mice were exposed to a different context B, which they associate with the fear-conditioning. The one day of training was sufficient to drive ChR2-EYFP expression, because of enhanced synaptic activity during learning. The critical part of the experiment comes last. Following training, the mice are put back into context A, an environment that they were habituated to and that they do not associate with noxious stimuli and fear. They are 'neutral' to this environment. Furthermore, they are put back on Dox. Exposure to blue light alone, which activates ChR2, now elicits freezing behavior— just as if the mice were exposed to the CS stimulus (presentation of a tone), expecting the electric foot shock, the US signal that they experienced during training. This result tells us that reactivation of those neurons that were also active during learning was sufficient to initiate the learned behavior, that is, to recall that memory. After training, about 6% of neurons in the mouse dentate gyrus had been labeled with ChR2-EYFP, comprising the component of the memory engram that was detectable in the microscopic, two-dimensional field of view.[5] Thus, as one could expect, the engram is comprised of a relatively small group of neurons, at least in the dentate gyrus. Of note, these neurons did not show clustering, but were widely distributed.

To be sure, this experiment only allows for an approximation, not for quantitative accuracy. This starts with the observation that *c-fos* activation is not restricted to synapses participating in the learning process. *c-fos* activation may result from synaptic activity not related to learning (*false positives*). Likewise, learning-related activity may not cause sufficient *c-fos* activation to detect the resulting fluorescent label (*false negatives*). The same lack of precision can be predicted for reactivation: too many or too few neurons might be activated, which may a cause a significant deviation from the true engram. Nevertheless, the arrival of the optogenetic toolbox has allowed neuroscientists to zoom in on memory engrams and to prove the very concept that memory is stored in the synchronous activation of connected cell populations. As engram research has just begun, the very concept of the 'engram' raises multiple new questions. One key

---

[5] A learned behavioral response to a fear-evoking stimulus is crucial for the animal's survival. It can be assumed that other types of memories are presented by even smaller, and more spatially restricted groups of neurons.

question is whether there is only one engram for each memory, or whether there are several layers of distinct populations of neurons, each of which could be called an engram with a hierarchically superior one to which other populations project. For example, the identification of a fear-memory-related engram in the dentate gyrus does not exclude the existence of additional distinct engrams in hippocampal areas CA3 or CA1—each of which, when labeled and appropriately activated by light, could drive the same learned behavior on its own. Related to this is the question of whether we should define an engram as a population of neurons that ultimately represents the encoded information content, or whether we should select a wider definition, in which all populations of neurons that undergo learning-related changes become engrams. Both approaches are possible, but future research will guide us toward a theoretical concept that is most useful to reach descriptive precision. The 'engram' concept also does not need to be restricted to populations of neurons. Semon's original definition of engram includes any lasting physiological or morphological change that results from repeated stimulus presentation. A perfect example of a memory engram—when the term is used as originally intended to describe a persistent 'engraving' of a memory trace—is provided by changes in the density or shape of dendritic spines. This spine plasticity concept primarily originated from the work of Tobias Bonhoeffer at the Max-Planck-Institute for Neurobiology in Munich (Engert and Bonhoeffer, 1999; Hofer et al., 2009; see also Trachtenberg et al., 2002). Such spine-based definition of a memory trace is evidently not in contradiction to one that focuses on activated neurons, as changes in spine and synapse function result in altered neuronal activation patterns. Instead, it highlights the importance of synaptic connectivity in the development and identity of neuronal engrams (see also Poo et al., 2016). In conclusion, it seems fair to say that we still have not been able to comprehensively describe engrams in their entirety, particularly not cortical engrams, and thus the search, initiated by Lashley, continues. However, we now have the technical toolbox at our disposal that is required to precisely describe engrams, and we have reason to be optimistic that we will be able to identify memory-specific engrams in the near future. Additional physiological parameters that I will describe in the next chapter will play a crucial role in engram function: nonsynaptic changes in membrane excitability.

# References

Abdou, K., Shehata, M., Choko, K., Nishizono, H., Matsuo, M., Muramatsu, S., and Inokuchi, K. (2018). Synapse-specific representation of the identity of overlapping memory engrams. *Science* 360, 1227–1231.

Bliss, T.V.P., and Lomo, T. (1973). Long-lasting potentiation of synaptic transmission in the dentate area of the anaesthetized rabbit following stimulation of the perforant path. *J. Physiol.* 232, 331–356.

Engert, F., and Bonhoeffer, T. (1999). Dendritic spine changes associated with hippocampal long-term synaptic plasticity. *Nature* 399, 66–70.

Hofer, S.B., Mrsic-Flogel, T.D., Bonhoeffer, T., and Hubener, M. (2009). Experience leaves a lasting structural trace in cortical circuits. *Nature* 457, 313–317.

Lashley, K. (1950). In search of the engram. *Symp. Soc. Exp. Biol.* 4, 454–482.

Liu, X., Ramirez, S., Pang, P.T., Puryear, C.B., Govindarajan, A., Deisseroth, K., and Tonegawa, S. (2012). Optogenetic stimulation of a hippocampal engram activates fear memory recall. *Nature* 484, 381–385.

Nabavi, S., Fox, R., Proulx, C.D., Lin, J.Y., Tsien, R.Y., and Malinow, R. (2014). Engineering a memory with LTD and LTP. *Nature* 511, 348–352.

Poo, M.M., *et al.* (2016). What is memory? The present state of the engram. *BMC Biology* 14, 40.

Reijmers, L.G., Perkins, B.L., Matsuo, N., and Mayford, M. (2007). Localization of a stable neural correlate of associative memory. *Science* 317, 1230–1233.

Titley, H.K., Brunel, N., and Hansel, C. (2017). Toward a neurocentric view of learning. *Neuron* 95, 19–32.

Trachtenberg, J.T., Chen, B.E., Knott, G.W., Feng, G., Sanes, J.R., Welker, E., and Svoboda, K. (2002). Long-term in vivo imaging of experience-dependent synaptic plasticity in adult cortex. *Nature* 420, 788–794.

# 8 | Limitations: The Non-Synaptic Plasticity Component[1]

Synapses convey information from one neuron to another. In other words, *what* the activity of a neuron stands for—that is, what it encodes—is entirely determined by the specific set of synaptic inputs that it receives. The information encoded can be about the state of the external world (sensory input, e.g. of visual or auditory nature) or about internal states (e.g. from muscle spindles, informing the brain about the degree of muscle stretch). The information can also represent the activity of other neurons, possibly reflecting internally generated cognitive concepts such as theory of mind.[2] The transfer of such information content from one cell to another is an exclusive privilege of synapses. Therefore, the establishment and strengthening of synapses during development, and the scaling of synaptic weights in the adult brain, are core features of learning. This, however, does not mean that synaptic plasticity is the *sole* cellular contributor to learning.

An experimental observation that has long called exclusively synaptic theories of learning into question is that when recordings are made from neurons after learning has taken place, these neurons show enhanced excitability (Disterhoft *et al.*, 1986; Moyer *et al.*, 1996; Schreurs *et al.*, 1998). These studies were done in slices prepared from the intact brain, subsequent to classical conditioning training (ex vivo). Injection of depolarizing current pulses into the

---

[1] This chapter is in parts adapted from Titley *et al.* (2017).
[2] Theory of mind describes the ability of the brain to think about mental states of the self and others.

soma evokes spike firing. The number of evoked spikes, the spike frequency, or the threshold of spike firing is taken as a measure of excitability. This type of experiment does not test synaptic weight. It exclusively tests cell-autonomous membrane excitability and its changes after learning. This nonsynaptic plasticity is also called 'intrinsic plasticity', because the change is confined to one cell only. Intrinsic plasticity does not strictly require the context of a behavioral learning situation. It can also be artificially triggered by tetanic activation of the soma. Examples are shown in Figure 8.1. Here,

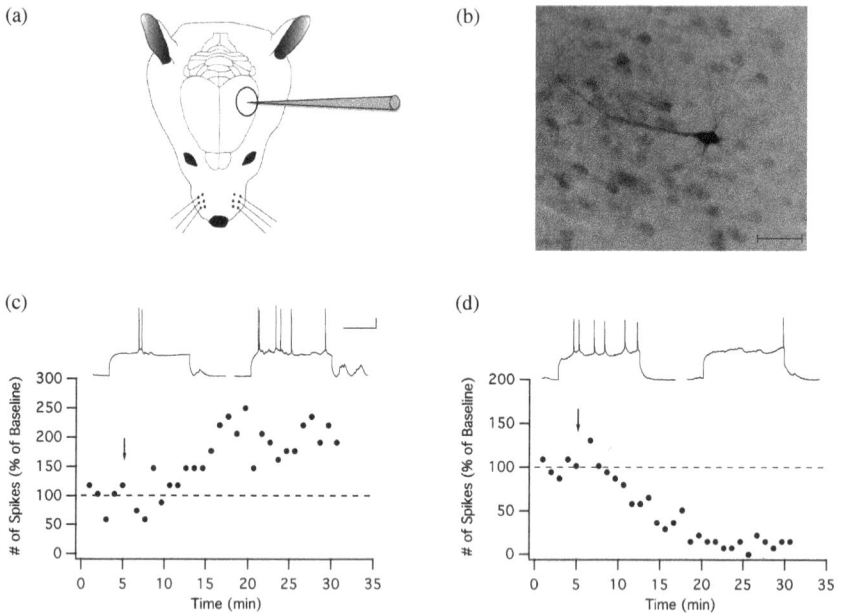

**Figure 8.1:** Intrinsic plasticity in the primary somatosensory cortex in vivo. (a) Recording configuration. Whole-cell patch-clamp recordings are performed from layer 2/3 pyramidal neurons in the barrel cortex of anesthetized rats. (b) Neurons are labeled with neurobiotin for histological identification subsequent to the recordings. The picture shows a neurobiotin-labeled layer 2/3 pyramidal neuron. Scale bar, 50 μm. (c and d) Example recordings illustrating the bidirectionality of intrinsic excitability changes. Intrinsic plasticity is triggered by repeated injection of depolarizing current pulses into the soma (5 Hz for 8 s). (c) In this example recording, tetanization resulted in an increase in excitability, as measured by the number of spikes evoked by test pulses. (d) Example of a neuron in which the same tetanization protocol resulted in a depression of the spike count. Scale bars, 200 ms/20 mV. Arrows indicate the time of tetanization. This figure is taken from Titley *et al.*, 2017.

depolarizing currents were injected at 5Hz for only a few seconds into the somata of cortical layer 2/3 pyramidal neurons of a rat under anesthesia. This tetanic current injection differs from the current injection in the test periods before and after tetanization. During the test periods, individual pulses are injected every 20 or 30 s to measure the excitability state. The goal is to avoid excitation of synapses and to exclusively activate the soma, which is in close proximity to the spike initiation site in the axon initial segment. This manipulation does not always result in the potentiation of spike firing. In fact, a case of a reduction in spike firing is shown in Figure 8.1d. In vivo recordings by Stephane Charpier at Sorbonne University in Paris have similarly shown that somatic current injection into layer 5 pyramidal neurons of anesthetized rats causes either a potentiation or a reduction of excitability, with a prevalence for a potentiating effect (Paz *et al.*, 2009; Mahon and Charpier, 2012). These findings show that intrinsic plasticity is a result of activity fed into neural circuits, and that neurons readily upregulate their excitability when being engaged in such activation. In recordings from layer 5 pyramidal neurons in vitro, it had previously been shown that the potentiation of excitability is mediated by a downregulation of small-conductance, calcium-activated SK-type $K^+$ channels (Sourdet *et al.*, 2003). A recent study from my own laboratory at the University of Chicago shows that in layer 2/3 pyramidal neurons, this SK channel-dependent intrinsic plasticity is facilitated by the activation of muscarinic acetylcholine receptors (Gill and Hansel, 2020). This suggests that attentional mechanisms, via cholinergic signaling, might be able to shift the probability from depression toward potentiation, ensuring that neurons enhance their excitability when they are activated in a behaviorally relevant context.

The demonstration of a functional role of such intrinsic plasticity in learning can be deducted from optogenetic studies, such as those described in the previous chapter. Remember that in work from Susumu Tonegawa and his team at MIT, it was possible to initiate fear memory recall by the light-induced activation of channelrhodopsin 2 (ChR2), which was selectively expressed under control of the *c-fos* promoter, and thus restricted to engram neurons (Liu *et al.*, 2012). The photoactivation of channelrhodopsin opens nonspecific cation channels that depolarize these neurons and initiates spike firing. Thus, in these studies, light-triggered memory recall rests on the channelrhodopsin-mediated depolarization that brings the cell closer to spike threshold. This is an important finding, as it shows that the need for a sufficiently strong synaptic drive can be bypassed by any means that make the cell fire action potentials. A related observation has been made in hippocampal place cells, which preferentially respond when an animal reaches a specific location in space. Albert Lee and his colleagues at Janelia Farm showed that some hippocampal CA1 pyramidal neurons are initially unresponsive to the local environment; they do not fire spikes when a rodent explores any particular region within a given environment. However, they can be activated and show emergent place cell properties when the excitability of the neuron is experimentally enhanced by the injection of depolarizing currents via a patch pipette in the whole-cell patch-clamp configuration (Lee *et al.*, 2012). In these examples, artificial depolarization—via photo-stimulation of channelrhodopsin or via patch-clamp—substitutes for depolarization by synaptic activation. The same effect, however, should also result when synaptic drive is kept constant, but membrane excitability is enhanced, such as in the activity-dependent downregulation of SK channels (Figure 8.2). Therefore, two scenarios are plausible to achieve enhanced spike firing upon synaptic activation: (a) enhanced synaptic drive, but unaltered membrane excitability, or (b) unaltered synaptic drive, but enhanced excitability.

Remarkably, evidence for an involvement of excitability can be found in the Tonegawa studies. When this group repeated the previously described experiments (Liu *et al.*, 2012), but in the presence of the protein synthesis inhibitor anisomycin, which blocks the late phase of LTP, photoactivation caused freezing at a rate that was indistinguishable from controls (Ryan *et al.*, 2015). Context-dependent freezing also took place, although at a reduced rate. Of note, anisomycin treatment prevented the strengthening of synaptic weights and the accompanying increase in spine density that is otherwise observed in the absence of anisomycin. Instead, an increase in intrinsic excitability was observed in engram cells (Ryan *et al.*, 2015; note

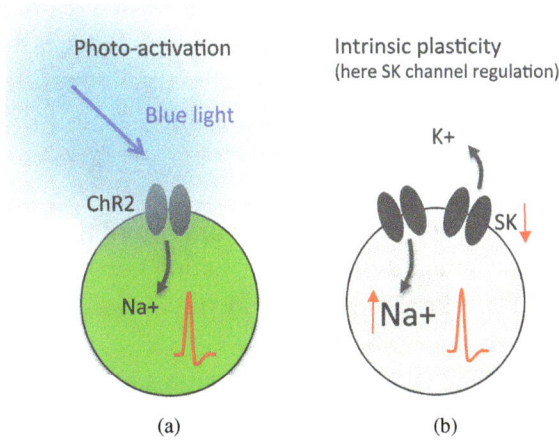

**Figure 8.2:** Depolarization toward spike threshold activates engram cells. (a) Depolarization via photo-activation. (b) Intrinsic plasticity results from a cell-autonomous modulation that similarly changes the probability of action potential generation. In the example shown here, SK channels are down-regulated to enhance membrane excitability. This figure is adapted from Titley *et al.*, 2017.

that SK channel-mediated plasticity is not sensitive to anisomycin, see Yamamoto *et al.*, 2019). This finding demonstrates that even in the absence of LTP, this type of learning can take place, albeit less efficiently, and that intrinsic plasticity may indeed be the underlying cellular mechanism.

These observations—enhanced membrane excitability promotes engram integration of neurons in the absence of synaptic potentiation, and such an increase in excitability is found in neurons after behavioral learning tasks have been completed—support the hypothesis that synaptic plasticity is not the only cellular learning correlate that is essential for all aspects of memory. In fact, it has previously been pointed out that properties of LTP do not match crucial properties of learning. For example, learning can result from single experiences, whereas LTP typically requires repetitive stimulation. (For this and other critiques of the synaptic learning theory, see Gallistel and Matzel, 2013; Gallistel and Balsam, 2014) In Chapter IX, I will offer an explanation of how single-experience learning might work. But let me first reiterate an extended learning hypothesis that I have proposed together with Heather Titley and Nicolas Brunel (Titley *et al.*, 2017), in which we assign a more broadly defined role in establishing connectivity maps to synapses, and add plasticity of intrinsic excitability as a mechanism for engram integration:

(1)  The decisive factor in memory engram formation and recall is the activation/integration of participating neurons (engram cells). Two plasticity processes are critically involved: synaptic and intrinsic plasticity.

(2)  Intrinsic plasticity sets an amplification factor that enhances or lowers *synaptic penetrance* and defines the neuron's presence within a memory engram. Here, we introduce the term *synaptic penetrance* in analogy to genetic penetrance to describe that synaptic weight changes (as in LTP) do not always result in enhanced spike firing. Yet other factors, such as intrinsic amplification, do provide important codeterminants. Intrinsic plasticity alone can, under some conditions, mediate engram cell integration, based on preexisting but unaltered synaptic connectivity.

(3)  Synapses play three fundamental roles in learning: they convey the specific information contents and input patterns that are to be memorized; their plasticity shapes connectivity maps by establishing connection patterns and by assigning synaptic weights; and their activity triggers intrinsic plasticity (induction phase) and drives the (re)activation of memory engrams, albeit without the need for accompanying changes in synaptic weight.

(4)  From the mentioned points, it follows that learned information is represented in two different ways in memory engrams: first, by the synaptic inputs that convey information to a neuron and collectively determine the coding identity of this neuron; and second, by the neuron itself, whose response threshold and activation characteristics determine its effect on target circuits and the representation weight of the information it encodes.

Key predictions of this extended learning hypothesis have now been experimentally confirmed by the group of Rafael Yuste at Columbia University (Alejandre-Garcia *et al.*, 2020). These researchers used a photo-stimulation protocol consisting of repetitive optogenetic activation to generate ensembles of neurons in the primary visual cortex of awake mice. These 'photo-ensembles' show spontaneous co-activity and can be re-activated by stimulation of individual member cells over consecutive days (Carrillo-Reid *et al.*, 2016). When such photo-ensembles were examined in closer detail using whole-cell patch-clamp recordings from layer 2/3 pyramidal neurons in the mouse primary visual cortex in vitro, Yuste and his team observed to their surprise that synaptic inputs to member neurons were unaltered, but the intrinsic excitability of these cells was enhanced. The observed alterations included a reduction in spike threshold and an increase in the number of spikes and spike frequency upon injection of current pulses (Alejandre-Garcia *et al.*, 2020). These important findings show that changes in membrane excitability alone, without changes in synaptic weight, may integrate neurons into ensembles and engrams, thus verifying central elements of our hypothesis.

The hypothesis maintains a central position for synaptic plasticity in information storage, but it departs from the classical view that LTP automatically enhances the probability for spike firing. The critical term chosen here is 'synaptic penetrance.' EPSPs that are locally generated in the dendrite propagate toward the soma ('forward propagation'), where the level of depolarization determines whether or not the spike threshold is reached and spike firing is initiated.[3] Using triple-patch recordings from layer 5 pyramidal neurons of the adult rat, Matthew Larkum, then a postdoctoral fellow with Bert Sakmann at the Max-Planck-Institute for Biomedical Research in Heidelberg, has demonstrated that distally generated dendritic potentials (patch pipette 1) vary dramatically in their amplitude at more proximal locations (patch pipette 2) and the soma (patch pipette 3): some are

---

[3] Ultimately, the site for action potential initiation is the axon initial segment, in close proximity to the soma.

strongly attenuated, some stay the same, and some are further amplified (Larkum *et al.*, 2001). The efficiency of forward propagation did not depend on further synaptic input, but likely on the state of nonlinearities—such as voltage- or calcium-dependent $K^+$ conductances, which we now know possess plasticity properties on their own.

The difficulty for locally evoked EPSPs to reach spike threshold is further illustrated by the drop in EPSP amplitudes from their origins in dendritic spines all the way to the soma. Using two-photon glutamate uncaging, and simultaneous two-photon voltage-sensitive dye recordings from cortical layer 5 pyramidal cells, a recent study measured spine potentials in the range of 6.5–30.8 mV (average, 13.0 mV) that evoked average somatic EPSPs of 0.59 mV (Acker *et al.*, 2016; for similar results, see Bloodgood *et al.*, 2009; Palmer and Stuart, 2009; Harnett *et al.*, 2012; Popovic *et al.*, 2015).

These low amplitudes of spine potentials arriving at the soma must be interpreted in the context of the distance from resting membrane potential to spike threshold. In layer 2/3 pyramidal neurons of the rat primary somatosensory cortex (here barrel cortex), the resting potential was found to be 15–40 mV below spike threshold. The average compound EPSP (defined as the response to stimulation of the principal whisker, regardless of the number of synapses involved) has an amplitude of 9.1 mV, resulting in an evoked spike rate of 0.031 action potentials per stimulus (Brecht *et al.*, 2003). In layer 5 pyramidal neurons, the voltage difference to spike threshold was 20.9 mV, with an average compound EPSP amplitude of 5.0 mV, resulting in a spike rate of 0.12 action potentials per stimulus (Manns *et al.*, 2004). In light of these results, previous findings make sense that described the phenomenon of 'silent neurons' in various cortical areas, neurons that do not respond with spike firing to the presentation of stimulus modalities that they should be 'hard-wired' for (Margrie *et al.*, 2002). Notably, this phenomenon is not found in all cortical areas. For example, visual cortices have been described to work under a 'high-input regime' (Shadlen and Newsome, 1998) that results from a bombardment by synaptic input. Reliable spike firing in response to visual stimuli has also been found in patch-clamp recordings from the cat primary visual cortex (Priebe and Ferster, 2005), which makes it unlikely that these different observations result from differences in the recording technique used. It is conceivable that pyramidal cells in different cortical subareas adjust their excitability to optimize neuronal function as either an 'event detector' (responses to whisker stimulation in barrel cortex: low excitability) or 'spike rate modulators' (variation of spike rate depending on visual stimuli in the visual cortex: high excitability), but this idea has not been experimentally tested.

The excitability of neuronal membranes can be modified by a variety of voltage-gated $Na^+$ and $Ca^{2+}$ conductances as well as voltage- or calcium-gated $K^+$ conductances. Of particular, but not exclusive, interest in a learning context are small-conductance, calcium-activated SK-type $K^+$ conductances. SK channels are exclusively activated by calcium, with an $EC_{50}$ for their calcium sensitivity of 0.3–0.5 µM (Xia et al., 1998).[4] SK conductances open fast, but decay relatively slowly, over a period of about 200 ms. This causes a medium-slow hyperpolarization following calcium influx, the so-called post-burst AHP (Stocker et al., 1999). Modulation of SK conductances therefore regulates excitability patterns, particularly following spike bursts. This effect in cerebellar Purkinje cells—in response to climbing fiber activity—is illustrated in Figure 8.3. Purkinje cells are spontaneously active. Climbing fiber stimulation causes a so-called complex spike that is followed by a spike pause. Bath application of the SK channel blocker apamin reduces the duration of this pause and enhances spike firing (Figure 8.3a). The pause results from an AHP that can be inspected in isolation when spontaneous spike firing is suppressed by the injection of a negative bias current. Figure 8.3b shows how this AHP is blocked when a calcium chelator (BAPTA) diffuses from the patch pipette into the cell—a result of preventing the calcium-driven activation of SK channels. The observation that these channels show plasticity, that is, they can be modulated in a lasting and activity-dependent way (Sourdet et al.,

---

[4]The $EC_{50}$ value describes the concentration at which a molecule or substance induces a half-maximal response.

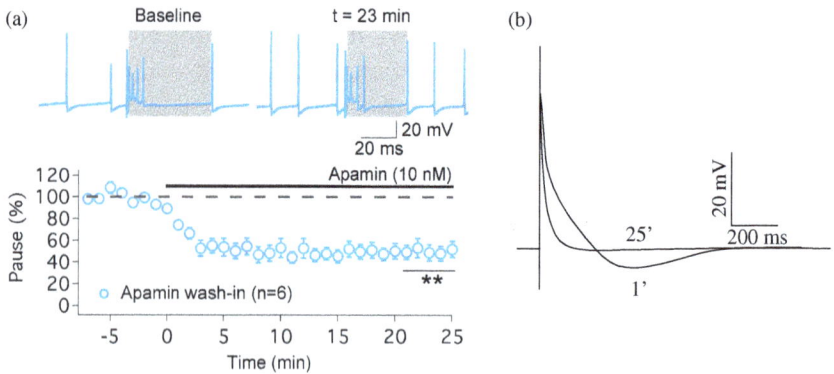

**Figure 8.3:** SK channels modulate excitability via their role in the post-burst afterhyperpolarization (AHP). (a) In whole-cell patch-clamp recordings from cerebellar Purkinje cells, activation of the climbing fiber input causes a complex spike (beginning of shaded area) that is followed by a pause. Application of the SK channel inhibitor apamin reduces the pause and enhances the overall spike rate. (b) When spontaneous spike firing is prevented by the injection of a negative bias current, the AHP underlying the pause becomes apparent. When the calcium chelator BAPTA is added to the patch pipette and diffuses into the cell, the calcium-sensitivity of the AHP conductance shows (here after a diffusion period of 25 min) and the AHP disappears. Panel A is taken from Grasselli *et al.* (2016). Panel B is taken from Schmolesky *et al.* (2005).

2003; Lin *et al.*, 2008; Belmeguenai *et al.*, 2010), makes them good candidates to regulate excitability and engram integration during learning processes.

How can we demonstrate that intrinsic plasticity—and here specifically SK channel plasticity—plays an essential role in learning? My laboratory at the University of Chicago has undertaken the task to provide evidence for such a role. Using cerebellar Purkinje cells and their role in motor learning as an example, we approached this problem by addressing three different questions:

1) Is the membrane excitability of Purkinje cells enhanced after training?
2) Is SK channel plasticity essential for the proper execution of a motor learning task?

3)  What is the cellular mechanism by which SK channel plasticity contributes to learning?

We selected delay eyeblink conditioning (EBC) as the motor learning task for these experiments. This type of cerebellum-dependent motor learning was introduced in chapter 5. Figure 5.4 illustrates how the conditioned stimulus (CS)- and unconditioned stimulus (US)-related information converges via parallel fiber and climbing fiber inputs onto Purkinje cells. These, in turn, regulate activity in the cerebellar nuclei to control eyelid closure. In collaboration with the group of John Disterhoft at Northwestern University, we addressed the first question and performed ex vivo recordings from Purkinje cells in mice that had undergone conditioning and, as a control, from mice that were pseudo-conditioned (i.e. the CS and US stimuli were presented in random order). We only recorded from Purkinje cells from an area (primary fissure near lobules V/HVI) that has been shown to be the microzone involved with the control of eyelid closure in mice (Heiney et al., 2014). In this study, we indeed found that the excitability of Purkinje cells from conditioned mice was enhanced as compared to control mice (Titley et al., 2020). This excitability increase was evident in the complex spike waveform by both a reduction in the AHP amplitude and an increase in the number of spike components. We also found a reduction in the AHP following parallel fiber bursts, which is shown in Figure 8.4c. Moreover, in slices from conditioned mice, we could not induce intrinsic plasticity any longer (which was still possible in recordings from control mice). The interpretation is that behavioral conditioning already utilized increases in excitability, resulting in an occlusion of further plasticity of membrane excitability.

We addressed the second question in collaboration with Chris De Zeeuw and Martijn Schonewille at the Erasmus Medical Center in Rotterdam (Chris is also a director of the Netherlands Institute for Neuroscience that is operated by the Royal Dutch Academy of Sciences in Amsterdam). Using mice with a specific knockout of SK2 channels in Purkinje cells (L7-SK2 mice; Purkinje cells only express the SK2 isoform, see Cingolani et al., 2002), we were able to show

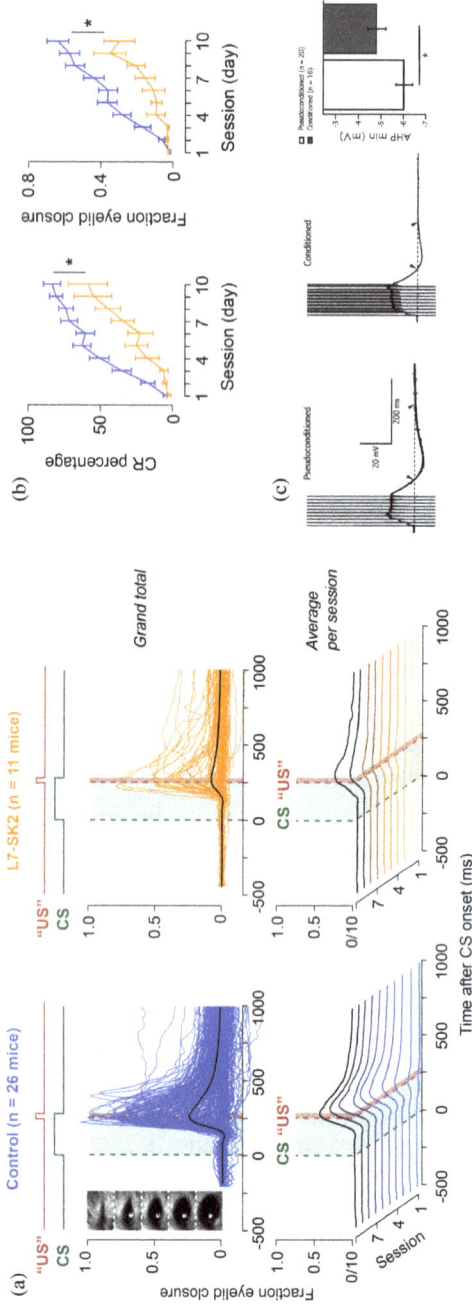

**Figure 8.4:** Intrinsic plasticity plays an essential role in motor learning. (a) Top: Eyeblink traces per mouse and total averaged eyeblink traces (black lines) recorded from control and L7-SK2 mice. The picture frames on the left are taken from a video illustrating eyelid closure ranging from 0 (fully open) to 1 (fully closed). Bottom: Waterfall plots showing the averaged eyeblink traces per group and per session. (b) L7-SK2 mice show a lower CR percentage (left) and lower fraction of the total eyelid closure (right; FEC or amplitude of eyelid closure) over the 10 training sessions. (c) In ex vivo recordings from eyeblink conditioned and pseudoconditioned mice, a reduction in the AHP amplitude (parallel fiber burst) indicates that conditioning resulted in increased Purkinje cell excitability. Panels (a) and (b) are taken from Grasselli et al. (2020). Panel (c) is taken from Titley et al. (2020).

that their EBC was impaired (Figure 8.4A + B) as evident in a reduced slope of the learning curve as well as a lower amplitude of eyelid movement (i.e. eyelid closure was less complete than in control mice). In recordings from Purkinje cells in slice preparations, we found that Purkinje cell intrinsic plasticity was absent in L7-SK2 mice, but synaptic plasticity was intact (Grasselli *et al.*, 2020). These observations allow us to conclude that SK2-dependent intrinsic plasticity can be separated in its functional consequences from synaptic plasticity, and that it is essential for the proper execution of this motor learning task.

We have previously established that intrinsic plasticity takes place in learning, and that it is indeed essential to ensure the completion of memory formation. But what is the exact role of excitability changes if information storage itself—a core feature of learning—is achieved by synaptic plasticity alone? To approach this problem, we returned to the observation made by Bert Sakmann's laboratory in neocortical pyramidal neurons that the EPSP amplitude measured in the proximal dendrite does not reflect well its amplitude at its site of origin in the distal dendrite (Larkum *et al.*, 2001). We began with this finding, and asked whether or not—in cerebellar Purkinje cells—the EPSP amplitude is a good predictor of spike output. In these experiments, we used two patch pipettes to record from Purkinje cells: one placed on the soma, and a second one on the dendrite, at a distance of about 100 μm from the soma (Figure 8.5a). We now applied a series of five stimuli to the parallel fiber input, resulting in an EPSP train of increasing amplitude (Figure 8.5b). As demonstrated here, the repeated injection of depolarizing current pulses (tetanization) enhanced the response amplitude and caused the firing of an action potential on EPSP 5. Application of the SK channel blocker apamin similarly enhanced responsiveness and spike firing. To determine the relationship between EPSP amplitude and spike firing, we focused our analysis on EPSP 5 only. The spike output was categorized into four groups: no spike, one spike, a weak burst (2 spikes) and a strong burst (3–6 spikes). We found that the amplitude of the dendritically recorded EPSP is a poor predictor of the spike output. There was a difference in EPSP amplitude between the no spike groups and all

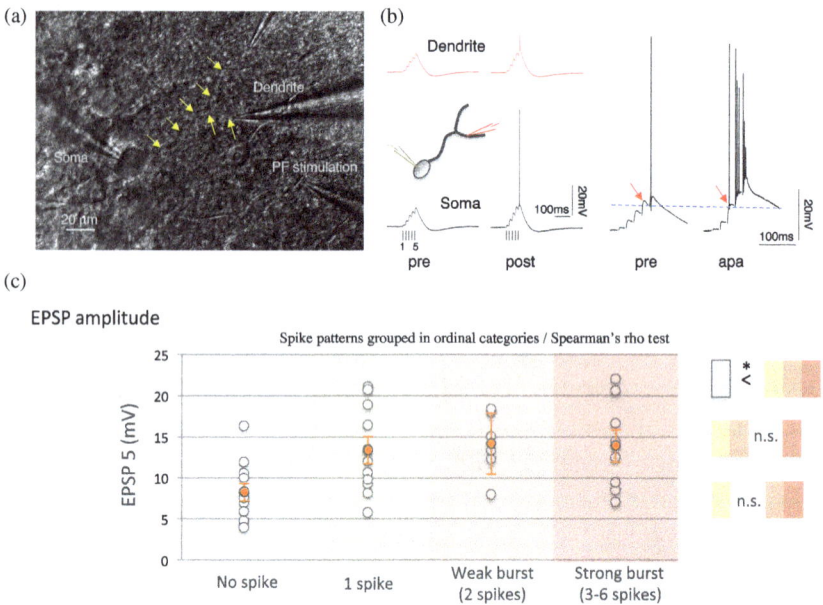

**Figure 8.5:** The dendritic EPSP amplitude is a poor predictor of the spike output. (a) Image illustrating the somato-dendritic recording configuration. Glass pipettes are used for recordings from the Purkinje cell soma (left) and the dendrite (right), as well as for parallel fiber stimulation (lower right). The yellow arrows point out the course of the primary dendrite. (b) Left: These recordings from the dendrite (top) and soma (bottom) show the emergence of spike firing on top of EPSP 5 after the application of the intrinsic plasticity protocol. Right: Example of a somatic recording showing that in the presence of apamin spikes were evoked by EPSPs (EPSP 3 in apamin; apa) that were smaller in amplitude relative to larger EPSPs recorded during the baseline that did not evoke spikes (EPSP 4 in pre; arrows). (c) Plot of the spike output versus the amplitude of EPSP 5. The spike output was categorized into four groups: no spike ($n = 11$), one spike ($n = 11$), weak burst (2 spikes; $n = 6$), and strong bursts (3–6 spikes; $n = 8$). These groups are color coded to illustrate the strength of the spike output. This figure is adapted from Ohtsuki and Hansel (2018).

spike groups, but whether one spike or six spikes were fired was entirely independent from the EPSP amplitude (Figure 8.5c). Instead, the spike output was well predicted by a combination of excitability measures, such as the AHP amplitude and the spike threshold (Ohtsuki and Hansel, 2018). This effect can also be seen in the example traces on the right in Figure 8.5b. In the presence of apamin, EPSP 3 evoked a single spike, although a larger EPSP 4 before apamin application did not (red arrows). These observations show that in Purkinje cells SK2 channels provide an intrinsic gate for the forward propagation of EPSPs that regulates the degree to which synaptic input can drive spike output. As we know that SK2 channels exhibit plasticity, these findings show that SK2 modulation/plasticity can regulate synaptic penetrance—that is, the degree to which synaptic input of a given weight has an impact on spike output—and with that, the participation of the entire neuron in neural ensembles or engrams. In the example of motor learning, this mechanism might be required to ensure that parallel fiber synapses, whose activity is required to enable beneficial motor programs, have a chance to contribute to the proper adjustment of spike output. In Purkinje cells, an individual additional spike will not have an impact on overall spike firing, as these cells are spontaneously active at frequencies up to 150 Hz (Häusser and Clark, 1997). This phenomenon explains the necessity for spike burst firing to override the background firing activity.[5]

# References

Acker, C.D., Hoyos, E., and Loew, L.M. (2016). EPSPs measured in proximal dendritic spines of cortical pyramidal neurons. *eNeuro* 3, e0050-15.2016.

Alejandre-Garcia, T., Kim, S., Perez-Ortega, J., and Yuste, R. (2020). Intrinsic excitability mechanisms of neuronal ensemble formation. bioRxiv doi. org/10.1101/2020.07.29.223966.

Belmeguenai, A., Hosy, E., Bengtsson, F., Pedroarena, C.M., Piochon, C., Teuling, E., He, Q., Ohtsuki, G., De Jeu, M.T., Elgersma, Y., De Zeeuw, C.I., Jörntell, H., and Hansel, C. (2010). Intrinsic plasticity complements long-term potentiation in parallel fiber input gain control in cerebellar Purkinje cells. *J. Neurosci.* 30, 13630–13643.

Bloodgood, B.L., Giessel, A.J., and Sabatini, B.L. (2009). Biphasic synaptic Ca influx arising from compartmentalized electrical signals in dendritic spines. *PLOS Biol.* 7, e1000190.

---

[5]In the recordings shown in Figure 8.5, this spontaneous background spiking was suppressed by the injection of negative bias currents.

Brecht, M., Roth, A., and Sakmann, B. (2003). Dynamic receptive fields of reconstructed pyramidal cells in layers 3 and 2 of rat somatosensory barrel cortex. *J. Physiol.* 553, 243–265.

Carrillo-Reid, L., Yang, W., Bando, Y., Peterka, D.S., and Yuste, R. (2016). Imprinting and recalling cortical ensembles. *Science* 353, 691–694.

Cingolani, L.A., Gymnopoulos, M., Boccaccio, A., Stocker, M., and Pedarzani, P. (2002). Developmental regulation of small-conductance Ca2+ -activated K+ channel expression and function in rat Purkinje neurons. *J. Neurosci.* 22, 4456–4467.

Disterhoft, J.F., Coulter, D.A., and Alkon, D.L. (1986). Conditioning-specific membrane changes of rabbit hippocampal neurons measured in vitro. *Proc. Natl. Acad. Sci. USA* 83, 2733–2737.

Gallistel, C.R., and Balsam, P.D. (2014). Time to rethink the neural mechanisms of learning and memory. *Neurobiol. Learn. Mem.* 108, 136–144.

Gallistel, C.R., and Matzel, L.D. (2013). The neuroscience of learning: beyond the Hebbian synapse. *Annu. Rev. Psychol.* 64, 169–200.

Gill, D.F., and Hansel, C. (2020). Muscarinic downregulation of SK2-type K+ conductances promotes intrinsic plasticity in L2/3 pyramidal neurons of the mouse primary somatosensory cortex. eNeuro 7, ENEURO.0453-19.2020

Grasselli, G., Boele, H.J., Titley, H.K., Bradford, N., van Beers, L., Jay, L., Beekhof, G.C., Busch, S.E., De Zeeuw, C.I., Schonewille, M., and Hansel, C. (2020). SK2 channels in cerebellar Purkinje cells contribute to excitability modulation in motor learning-specific memory traces. *PLOS Biology* 18, e3000596.

Grasselli, G., He, Q., Wan, V., Adelman, J.P., Ohtsuki, G., and Hansel, C. (2016). Activity-dependent plasticity of spike pauses in cerebellar Purkinje cells. *Cell Reports* 14, 2546–2553.

Harnett, M.T., Makara, J.K., Spruston, N., Kath, W.L., and Magee, J.C. (2012). Synaptic amplification by dendritic spines enhances input cooperativity. *Nature* 491, 599–602.

Heiney, S.A., Kim, J., Augustine, G.J., and Medina, J.F. (2014). Precise control of movement kinematics by optogenetic inhibition of Purkinje cell activity. *J. Neurosci.* 34, 2321–2330.

Larkum, M.E., Zhu, J.J., and Sakmann, B. (2001). Dendritic mechanisms underlying the coupling of the dendritic with the axonal action potential initiation zone of adult rat layer 5 pyramidal neurons. *J. Physiol.* 533.2, 447–466.

Lee, D., Lin, B.J., and Lee, A.K. (2012). Hippocampal place fields emerge upon single-cell manipulation of excitability during behavior. *Science* 337, 849–853.

Lin, M.T., Lujan, R., Watanabe, M., Adelman, J.P., and Maylie, J. (2008). SK2 channel plasticity contributes to LTP at Schaffer collateral-CA1 synapses. *Nat. Neurosci.* 11, 170–177.

Liu, X., Ramirez, S., Pang, P.T., Puryear, C.B., Govindarajan, A., Deisseroth, K., and Tonegawa, S. (2012). Optogenetic stimulation of a hippocampal engram activates fear memory recall. *Nature* 484, 381–385.

Mahon, S., and Charpier, S. (2012). Bidirectional plasticity of intrinsic excitability controls sensory inputs efficiency in layer 5 barrel cortex neurons *in vivo. J. Neurosci.* 32, 11377–11389.

Manns, I.D., Sakmann, B., and Brecht, M. (2004). Sub- and suprathreshold receptive field properties of pyramidal neurones in layers 5A and 5B of rat somatosensory barrel cortex. *J. Physiol.* 556, 601–622.

Margrie, T.W., Brecht, M., and Sakmann, B. (2002). In vivo, low-resistance, whole-cell recordings from neurons in the anaesthetized and awake mammalian brain. *Pflugers Arch.* 444, 491–498.

Moyer, J.R., Jr., Thompson, L.T., and Disterhoft, J.F. (1996). Trace eyeblink conditioning increases CA1 excitability in a transient and learning-specific manner. *J. Neurosci.* 16, 5536–5546.

Ohtsuki, G., and Hansel, C. (2018). Synaptic potential and plasticity of an SK2 channel gate regulate spike burst activity in cerebellar Purkinje cells. *iScience* 1, 49–54.

Palmer, L.M., and Stuart, G.J. (2009). Membrane potential changes in dendritic spines during action potentials and synaptic input. *J. Neurosci.* 29, 6897–6903.

Paz, J.T., Mahon, S., Tiret, P., Genet, S., Delord, B., and Charpier, S. (2009). Multiple forms of activity-dependent intrinsic plasticity in layer V cortical neurones *in vivo. J. Physiol.* 587, 3189–3205.

Popovic, M.A., Carnevale, N., Rozsa, B., and Zecevic, D. (2015). Electrical behavior of dendritic spines as revealed by voltage imaging. *Nat. Commun.* 6, 8436.

Priebe, N.J., and Ferster, D. (2005). Direction selectivity of excitation and inhibition in simple cells of the cat primary visual cortex. *Neuron* 45, 133–145.

Ryan, T.J., Roy, D.S., Pignatelli, M., Arons, A., and Tonegawa, S. (2015). Engram cells retain memory under retrograde amnesia. *Science* 348, 1007–1013.

Schmolesky, M.T., De Zeeuw, C.I., and Hansel, C. (2005). Climbing fiber synaptic plasticity and modifications in Purkinje cell excitability. *Prog. Brain Res.* 148, 81–94.

Schreurs, B.G., Gusev, P.A., Tomsic, D., Alkon, D.L., and Shi, T. (1998). Intracellular correlates of acquisition and long-term memory of classical conditioning in Purkinje cell dendrites in slices of rabbit cerebellar lobule HVI. *J. Neurosci.* 18, 5498–5507.

Shadlen, M.N., and Newsome, W.T. (1998). The variable discharge of cortical neurons: implications for connectivity, computation, and information coding. *J. Neurosci.* 18, 3870–3896.

Sourdet, V., Russier, M., Daoudal, G., Ankri, N., and Debanne, D. (2003). Long-term enhancement of neuronal excitability and temporal fidelity mediated by metabotropic glutamate receptor subtype 5. *J. Neurosci.* 23, 10238–10248.

Stocker, M., Krause, M., and Pedarzani, P. (1999). An apamin-sensitive $Ca^{2+}$-activated $K^+$ current in hippocampal pyramidal neurons. *Proc. Natl. Acad. Sci. USA* 96, 4662–4667.

Titley, H.K., Brunel, N., and Hansel, C. (2017). Toward a neurocentric view of learning. *Neuron* 95, 19–32.

Titley, H.K., Watkins, G.V., Lin, C., Weiss, C., McCarthy, M., Disterhoft, J.F., and Hansel, C. (2020). Intrinsic excitability increase in cerebellar Purkinje cells following delay eyeblink conditioning in mice. *J. Neurosci.* 40, 2038–2046.

Xia, X.M., Fakler, B., Rivard, A., Wayman, G., Johnson-Pais, T., Keen, J.E., Ishii, T., Hischberg, B., Bond, C.T., Lutsenko, S., Maylie, J., and Adelman, J.P. (1998). Mechanism of calcium gating in small-conductance calcium-activated potassium channels. *Nature* 395, 503–507.

Yamamoto, M., Kim, M., Imai, H., Itakura, Y., and Ohtsuki, G. (2019). Microglia-triggered plasticity of intrinsic excitability modulates psychomotor behaviors in acute cerebellar inflammation. *Cell Reports* 28, 2923–2938.

# 9

# Ensemble Sequences, Associative Plasticity and the Learning of Causal Relationships[1]

## Intrinsic Plasticity and Learned Feature Generalization

Intrinsic plasticity complements synaptic weight changes by controlling forward propagation of dendritic potentials toward the soma (the cell body). As we have seen in Chapter VIII, downregulation of SK conductances creates a 'high-excitability state' that enables spike firing to be driven by otherwise inefficient synaptic excitation. There are, however, additional implications of intrinsic plasticity that influence features of cellular learning. First, intrinsic plasticity will determine to what degree neurons get integrated into neural ensembles, which can be conceptually imagined as populations of synchronously active neurons (Figure 9.1; see also Singer, 1995). In this view, intrinsic plasticity adds an intrinsic amplification factor that acts like a light dimmer, adjusting synaptic penetrance via two effects: by boosting the response amplitude and by lowering the spike threshold. For example, in recordings from layer 5 pyramidal neurons in the barrel cortex of anesthetized rats, it was found that intrinsic potentiation and depression, respectively, modulate the spike firing threshold by about 1–2 mV, suggesting a total dynamic range of up to 4 mV (Mahon and Charpier, 2012).

---

[1] This chapter is in parts adapted from Titley et al. (2017).

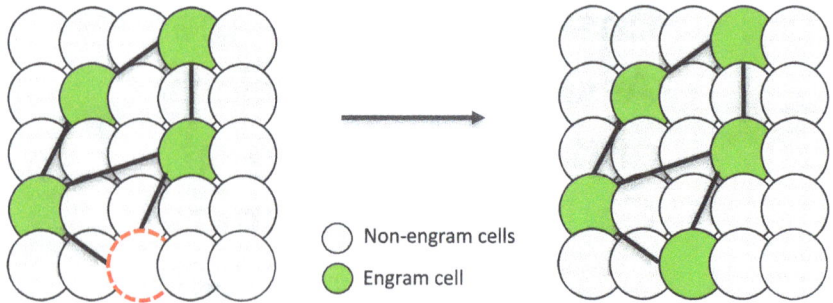

**Figure 9.1:** Integration of neurons into ensembles or memory engrams may result from increases in their membrane excitability. Here, a neuron that is synaptically connected with engram cells but does not engage (red stippled line; left) becomes integrated (right), for example, in response to synaptic drive. No changes in synaptic connectivity weights are required for this integration step.

Second, intrinsic plasticity, when it takes place close to the soma and thus near the site of action potential generation, will similarly affect all synaptic inputs onto a neuron. This is because all synaptic responses will be further amplified and will more easily reach spike threshold. (Likewise, intrinsic plasticity can reduce excitability and make it more difficult to reach the spike threshold.) Although intrinsic plasticity might not provide the initial steps of information encoding, which happen at the synapses, it nevertheless becomes an integral part of the learning machinery by enabling synapses to drive action potential firing and neurons to participate in memory engrams. As a consequence of this neuron-wide scaling effect, all synaptic inputs, independent of their recent potentiation or depression history or their current synaptic weight, share the same amplification fate. Therefore, they will be up- and downregulated together. If a neuron's synaptic inputs are random and convey largely unrelated information (Figure 9.2, left), then *shared* intrinsic scaling might not have much of a purpose. Combinatorial encoding across several neurons will reveal what information was stored. However, if a neuron's synaptic input portfolio is dominated by one or a few sets of synaptic inputs (Figure 9.2, middle), or if the majority of synapses convey largely related information, then intrinsic plasticity and the shared amplification fate of its inputs will define a coding identity of the neuron, that

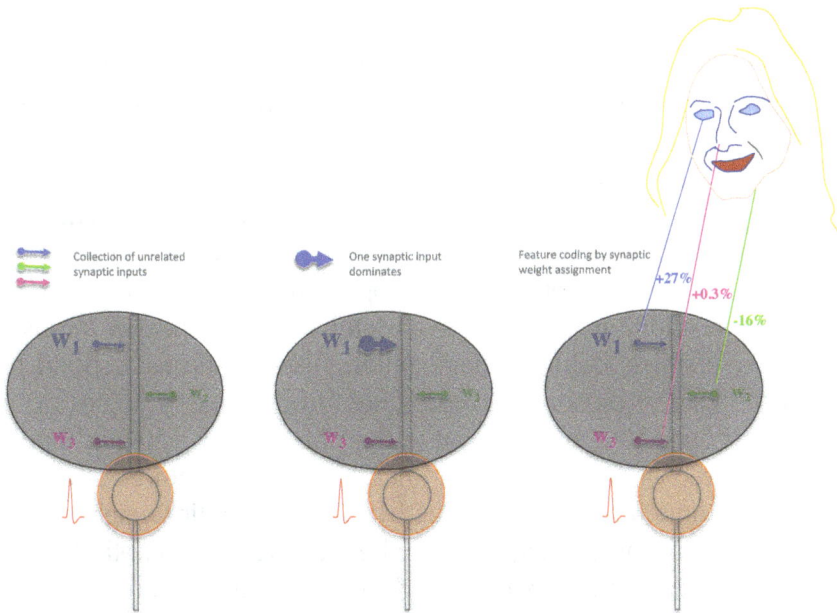

**Figure 9.2:** Intrinsic amplification is shared by synaptic inputs. Intrinsic plasticity affects the probability of spike firing (red action potential symbol; red circles indicate the somatic area where intrinsic amplification takes place). The grey circles around the dendritic input area indicate that synaptic responses, regardless of synaptic input weight W, are exposed to the same amplification factor. In the scheme on the left, synaptic inputs convey largely unrelated information. In the scheme in the middle, one input, or a cluster of inputs, dominates, giving the neuron a 'coding identity'. In the cell on the right, synaptic inputs work together to encode a complex feature, such as in concept cells coding for specific individual faces.

is, the activity of this neuron encodes specific information. In the extreme case—this is likely true for neurons that code for complex feature constellations such as face recognition cells—the many synapses that connect onto a target neuron encode information related to one specific object (Figure 9.2, right). Face recognition cells have been discovered in humans and nonhuman primates. These neurons respond selectively to images of specific individuals. Such experiments have been performed using pictures of prominent people, such as former US president Bill Clinton or the actress Jennifer Aniston (e.g. Quiroga *et al.*, 2005), and have therefore been coined

'Bill Clinton cells' or 'Jennifer Aniston cells' (Figure 9.2). If such neurons also respond to images that identify a person in an indirect or abstract way, such as showing the written name (in human studies), these cells qualify as 'concept cells', that encode conceptual representations of a given object (for a review, see Quiroga, 2012). Here, the advantages of intrinsic plasticity become obvious. It enables changes in the availability of specific memories without changes in synaptic weight ratios, which would disturb proper feature representation.

Face recognition cells provide a useful illustration of how intrinsic plasticity enables feature generalization. These neurons respond to the same face or person when presented from different angles or in different contexts. It is conceivable that the various synaptic inputs onto face recognition cells each encode different features of a person's face and represent them with unique synaptic weights. This scenario is illustrated in Figure 9.2, where changes in synaptic weight (in percentages) from a theoretical default state are presented. Other synapses onto the same neuron may encode additional features that might be more strongly weighted in a different set of pictures. As all these inputs share the same amplification fate via intrinsic plasticity, strong activity of such a neuron would represent a specific person regardless of presentation details. Such 'generalization' also occurs at a less specific level in associative learning, and intrinsic plasticity may well play a role here, too. After acquisition of a conditioned response (CR) and extinction to a specific conditioned stimulus (CS)-unconditioned stimulus (US) pair (e.g. tone and air puff), subsequent acquisition of a CR using a different CS (e.g. light) occurs faster (for a discussion, see Hansel et al., 2001). Because generalization can occur with various CS configurations in the first and second training sessions, this phenomenon cannot be readily explained by localized synaptic changes in isolation, but is likely related to widespread changes in intrinsic plasticity. In this scheme, synaptic plasticity links new information to a target neuron, thus stabilizing synapses that represent CS-related input. In the case of a 'Jennifer Aniston cell,' a new representation of the actress, for example, her written name instead of an image, could be presented and learned by the neuron (see Quiroga, 2012). Intrinsic plasticity would not 'connect' this new information to the neuron, but would adjust excitability, and with it the robustness of the conceptual representation of Jennifer Aniston, regardless of whether her written name or her image was shown. LTP on its own is able to connect subsets of synapses, each conveying different aspects of the concept 'Jennifer Aniston', to a target neuron. As a result, the target neuron represents generalized information about the actress. However, true generalization implies that 'Jennifer Aniston cells' can be brought to spike firing independent of any specific subset of synaptic input—which would give preference to strong driver inputs. In contrast, cell-autonomous intrinsic plasticity would enhance excitability, enabling both weak and strong inputs to drive spike firing. This step makes the activation process more 'democratic' and completes the generalization process.

## Effect Robustness and Duration

What are the different functions of synaptic and intrinsic plasticity in learning? As mentioned, synapses store information, but they may have less impact on the integration of neurons into memory engrams than originally anticipated. In contrast, intrinsic plasticity does not primarily store information, but is well suited to regulate engram participation. An additional parameter that influences overall function is effect duration. LTP and other types of synaptic plasticity are seen as lasting alterations in synaptic weight, capable of persisting for years (Abraham, 2003). In the hippocampus, intrinsic plasticity has been observed after trace eyeblink conditioning, another type of associative learning, but these changes return to baseline levels within seven days (Moyer et al., 1996; Thompson et al., 1996). In hippocampal dentate gyrus neurons, excitability is enhanced after fear conditioning, but excitability returns much faster to baseline, within a time frame of about 2 hrs (Pignatelli et al., 2019). Because of the transient nature of excitability changes in intrinsic plasticity, it has been argued that the excitability increase facilitates LTP and integrates neurons into engrams via a lasting strengthening of synaptic connectivity ('memory allocation hypothesis'; Thompson et al., 1996; Yiu et al., 2014; Cai et al., 2016; Rashid et al., 2016). Although this is a plausible scenario, intrinsic plasticity may play a more persistent role. In cerebellar Purkinje cells, enhanced excitability has been measured a month after eyeblink conditioning (Schreurs et al., 1998). Our own ex vivo study of excitability changes in Purkinje cells (Figure 8.4) has not focused on the question of effect duration, but we observed these changes 2 days after conditioning (Titley et al., 2020). Thus, intrinsic plasticity does not necessarily fade within minutes or hours. It is conceivable that intrinsic plasticity itself, without the need for further LTP, can integrate neurons into engrams over prolonged periods of time, at least in some types of neurons.

There is a crucial and relevant feature of intrinsic plasticity that I have not yet discussed. It appears that excitability changes are easier to trigger than synaptic weight changes. In other words, the induction threshold for intrinsic plasticity may be lower than that for synaptic plasticity. In slice experiments, LTP induction usually requires stimulation in the range of minutes, although there are types of LTP that can be triggered using activation protocols that last 15–20 seconds (Salin et al., 1996; Frick et al., 2004). Intrinsic plasticity on its own can be evoked with ultra-short activation protocols that have not been similarly reported for LTP induction. For example, Purkinje cell intrinsic plasticity results from synaptic or somatic (injection of depolarizing currents) activation periods as short as 3 sec (Belmeguenai et al., 2010; Ohtsuki et al., 2012). Comparing the outcome of a range of tetanization protocols on plasticity in dentate gyrus granule cells, it has recently been shown that intrinsic plasticity results from mild conditioning protocols. Conversely, synaptic plasticity emerges when stronger protocols are applied, suggesting that in general, intrinsic plasticity might have a lower induction threshold than synaptic plasticity (Lopez-Rojas et al., 2016). If excitability changes tend to be less stable than synaptic weight changes, but can also be initiated more easily, intrinsic plasticity may not fit well into the classical definition as either short-term or long-term plasticity. Instead, intrinsic plasticity may present a 'flickering state,' in which membrane excitability has the mechanistic potential to stay elevated for extended periods of time, but quickly transitions between amplification states based on the recent history of neural activity.

A consequence of a lower induction threshold for intrinsic plasticity compared with synaptic plasticity is that membrane excitability can change as a result of synaptic drive, without the simultaneous induction of LTP. This indicates that cellular plasticity can serve more functions in learning than previously anticipated. Imagine that you know about the actress Jennifer Aniston, but you have not seen one of her movies in years, and so she is not part of your portfolio of often-recalled memories. The concept 'Jennifer Aniston' is in your memory, but somewhere deep, not easily accessible. But the next time you see a Jennifer Aniston movie, you might think about it for a few days. The concept 'Jennifer Aniston' will be more actively present. There is a hierarchy of memories, and the cellular features of intrinsic plasticity provide a perfectly appropriate mechanism to control such hierarchy (Figure 9.3). There is no need to change synaptic weights when elevating memories to a higher position in the memory hierarchy. Intrinsic plasticity can do the job without further LTP events, and without the risk of changing the weight ratios of different synaptic inputs, and with it the risk that a concept cell might represent altered content. Similarly, intrinsic plasticity may underlie single-experience learning. It is hard to imagine that LTP results from exposure to an object or an experience after just a few seconds. However, if there are concept cells that represent a

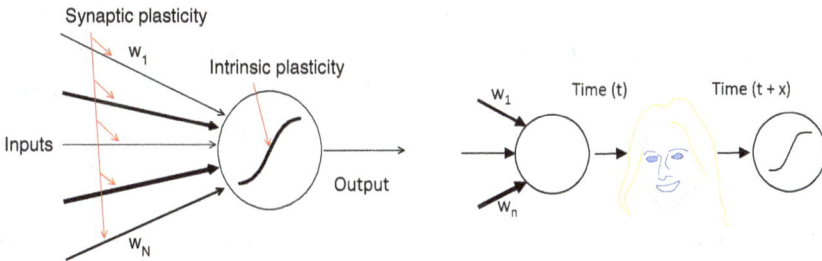

**Figure 9.3:** Intrinsic and synaptic plasticity can be separated in time. Left: sketch of a simplified single-neuron model showing synaptic weight adjustment by synaptic plasticity. Synaptic input weights are summed linearly and passed through a static nonlinearity. Loci of plasticity are indicated in red. Right: illustration of the hypothesis that the adjustment of synaptic weights (synaptic input map) and the plasticity of the intrinsic amplification factor can be separated in time. The example shown here depicts a face recognition cell ('Jennifer Aniston cell'). A time axis is indicated by the arrows to the left and right of the picture. Intrinsic plasticity, occurring at time $t + x$, may enhance engram representation without altering synaptic weight ratios, but solely based on pre-existing connectivity. This figure is adapted from Titley $et$ $al.$ (2017).

default generalization for an object or an experience, made possible by an existing synaptic connectivity map, then intrinsic plasticity could be triggered within the time frame of a typical single experience exposure. Subsequently, this could cause the integration of that concept cell into a wider memory engram. Excitability changes in behavioral learning can indeed be triggered after single trial exposure (McKay *et al.*, 2009). These potential functions of nonsynaptic plasticity illustrate how excitability changes always depend on processes that stabilize synapses and provide physical connectivity, but also how they might act on their own at certain times and contribute to crucial features of memory.

## Synaptic and Intrinsic Mechanisms Control the Probability of Engram Activation and Engram Sequences

### Intrinsic Contributions

Memory can be subdivided into chronologically distinct phases. An 'induction phase' can be separated from a 'maintenance phase', and 'storage mechanisms' can be separated from 'recall mechanisms'. It is remarkable that the scientific disciplines invested in the study of learning and memory—in particular, psychology and neuroscience—have yet to develop proper terminology, not to mention a conceptual framework, to describe the probability at which specific memories are recalled. This factor is not automatically covered by the study of recall itself. Research on recall typically focuses on mechanisms that enable recall, as opposed to mechanisms that would set probabilities to do so, that is, contribute to the occurrence of memory hierarchies. Engrams and, more broadly, all neural ensembles[2] are

---

[2]We have earlier in this book distinguished the terms 'engram' and 'ensemble' by assigning a memory context to engrams. This means that an engram stores memory content. Note, however, that all types of ensembles are shaped by memory mechanisms, as their composition depends on synaptic connectivity and the excitability of participating neurons.

probabilistically activated in sequences, in a process that is shaped by prior experience and cellular plasticity. It is likely that factors such as short-term plasticity and neuromodulator signaling play a role in determining which sequence is triggered at any given time. Synaptic connectivity is a prerequisite, as synaptic drive provides the only plausible external trigger to activate an ensemble or engram. The ultimate need for suprathreshold activation of participating engram cells, however, assigns a criticality to the cell-autonomous regulation of membrane excitability. This places intrinsic plasticity at the very top of cellular mechanisms that have the capacity to regulate the sequential activation of engrams. The observation that intrinsic plasticity—this certainly holds true for SK channel plasticity—is modulated by cholinergic signaling (Gill and Hansel, 2020) only strengthens this argumentation.

## Synaptic Contributions

The transition from the activation of one ensemble or engram to the next requires synaptic connectedness. Associative synaptic plasticity plays a role in establishing appropriate connections. A need for associative learning can be identified in sequential output, such as language. I will focus here on the learning of simple associations of the type 'A is associated with B', or 'A is causally related to B'. There is no animal behavior involved, as it would be in the two best-known experimental tests for associative learning: classical and operant conditioning. Let us consider a brief statement—spoken, written or thought—such as 'I love Paris in the spring' (1):

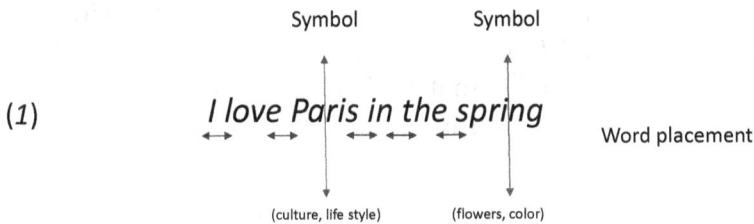

(1)

Symbol          Symbol

*I love Paris in the spring*          Word placement

(culture, life style)          (flowers, color)

There are two types of associations that have to be formed here. First, as pointed out by the German philosopher Ernst Cassirer in his

1923 book *Philosophy of Symbolic Forms*, words are symbols for the
content that they represent.[3] Thus, the word 'Paris' represents (sym-
bolizes) the French capital and also provides a symbol for our collec-
tive associations with this city name—sights, such as Montmarte,
Notre Dame, or the Eiffel Tower, but also restaurants, fashion and
lifestyle. The word 'spring' stands for the season, but also for collec-
tive associations, such as the opening of blossoms, warm temperature, and colors that replace the grey of winter days. The words 'Paris'
and 'spring' constitute symbolic representations. This point becomes
more obvious when considering that they can easily be replaced by
other symbols, including equivalent words in different languages. Let
us have a lock at the same sentence in two other languages that—like
English—belong to the group of West Germanic languages[4] and thus
have a similar sentence structure, German (*2*) and Dutch (*3*):

(2)   *Ich liebe Paris im Frühling*
(3)   *Ik hou van Parijs in het voorjaar*

The city name, 'Paris,' is the same in English and German, but
differs slightly in Dutch. In contrast, the words for 'Spring' are different in English, German (*Frühling,* or *Frühjahr*) and Dutch (*Voorjaar*),
and yet, they are all equally efficient symbols for the same content:
the season. These different words have meaning in the sense that (a)
they can be composed of sub-words that have meaning on their own
and (b) the words (or sub-words) are derived from words in older,
related languages. For example, the German word *Frühjahr* (which is
closely related to the Dutch *Voorjaar*) is composed of *Früh* (early) and
*Jahr* (year), thus meaning 'early in the year'—and therefore is perfectly descriptive. *Jahr* is derived from the Old High German word
*jar,* and *Früh* is derived from the Old High German word *fruoji.* (Old
High German was spoken in areas of Middle Europe between the

---

[3]Cassirer, E. (1923). *Philosophie der symbolischen Formen. Erster Teil: Die Sprache*
(Berlin: Bruno Cassirer).
[4]The West Germanic languages include German, English, and Dutch, as well as
Yiddish (derived from German) and Afrikaans (derived from Dutch).

years 750 to 1000.) Like many other words in the West Germanic languages, it is difficult to trace the origin of these words back any further, though like any language, they almost certainly have their origins in older languages. (Some modern artificial words are the exception). Despite such 'meaning' of words, Cassirer is right when arguing that words are primarily symbolic representations; they usually achieve meaning when content or an object is assigned to them. This is especially true of speakers of modern languages; they are typically unaware of the origin of most words, which they use largely symbolically. In addition, we could quickly learn the use of an entirely nondescriptive fantasy word, such as *Fojo*. Regardless of the origin of the word, we give meaning to it by learning the association of the word with the object that it describes. (This is similar to how we would memorize the name of our neighbor's new dog: it might be *Susie* or *Fojo*; names are more often than not nondescriptive and are learned by association alone.)

Second, correct *word placement* is dictated by the grammatical rules of a language, as well as by the need to generate logical word sequences or hierarchies that create meaning. It is possible that the syntactic structure of language does not have be learned, but that its principles are in part genetically inherited, as argued by the American linguist Noam Chomsky in his 1957 book *Syntactic Structures*, and later by the Canadian psychologist and linguist Steven Pinker in his 1994 book *The Language Instinct*. However, correct word placement also implies that words need to fit into a sentence, in the sense that, in combination with words that have been used at an earlier position in the same sentence or the sentence before that, they create meaning. This process relies on learning word associations, for example, in appropriate word pairs or established phrases.

A particularly intriguing aspect of cognition is our ability to generate prolonged sequences of thought—in the absence of external input. These thought sequences might be a series of sentences such as *I love Paris in the spring*, but they might also be based on music (tunes played in our heads) and mathematics (formula sequences). Blindfold chess is an example related to *thought* mathematical operations that have special meaning for me. (My father was an ardent chess player and occasionally played blindfold chess with my brother and me when we were teenagers. With awe, we listened to him telling us about blindfold chess games featuring the most brilliant grandmasters of chess, such as José Raúl Capablanca or Mikhail Botwinnik. The record in blindfold chess was set in 2016 by Timur Gareyev, who simultaneously played 48 blindfold chess games, of which he won 35). Thought language and music are built from prolonged association sequences. Mathematical operations (including blindfold chess) are based on sequential calculations that in part involve causal relationships. These types of thought processes all rely heavily on the use of symbols, and therefore require associative plasticity to form associations between symbol and object.

The meaning of language that is thought, rather than spoken or written, has been reflected upon by the German cultural critic and philosopher Walter Benjamin. In his 1916 essay *Über Sprache überhaupt und über die Sprache des Menschen*[5] (on language in general and on the language of man), Benjamin argues for the primacy of our ability to *think* language in defining us as intelligent beings. To express this idea, Benjamin used the phrase '*in* der Sprache.... nicht *durch* die Sprache' (*in* language, not *through* language) to emphasize that it was not primarily the communicative function of language (*through* language) that makes us intelligent beings: rather it is the fact that '*in* language' complex meaning can be encoded, even if only thought, and not communicated. It is obvious that communicated language is required for the development of speech and language, as well as for the development of societies. Nevertheless, Benjamin's essay makes an important point: *thought language* is a requirement for *thought itself*.

What are the neural circuits that support aspects of associative plasticity in language, and that therefore may also enable sequences of thought language? Many brain areas, including the neocortex, hippocampus, and cerebellum, support forms of associative learning (e.g. Hattori *et al.*, 2014). Any association of 'A' with 'B', such as a

---

[5]Benjamin, W. (1991). Über Sprache überhaupt und über die Sprache des Menschen, in *Gesammelte Schriften* (Frankfurt: Suhrkamp), pp. 140–157.

meaningful association between the words 'read' and 'book', can result from LTP at specific synapses. That is because synaptic activation representing the information content 'A' needs to facilitate the activation of synapses, neurons, or ensembles, whose activity represents 'B'. Theoretically, LTD can have this effect, too, if, for example, LTD at excitatory synapses onto inhibitory interneurons reduces the inhibition that pyramidal neurons receive. Thus, the cellular plasticity machinery that can support associative learning is not different from machineries for learning in general and should be available in many different brain circuits. Specializations, such as supervised 'error'-guided learning in the cerebellum, do exist, however, and I will discuss cerebellar learning in detail in the following. Before doing so, I would like to re-emphasize that the basic forms of synaptic plasticity, LTP and LTD, are capable of supporting associative learning in several brain areas. In language acquisition, in particular, it can be said with certainty that cognitive operations in the neocortex play an important role.

Here, though, I focus on properties of associative plasticity in the cerebellum, a brain structure in which associative synaptic learning has been studied in rich detail. Before turning to nonmotor aspects of language, I would like to note that the cerebellum plays a known, but poorly understood role in sequence generation—and in learning causal relationships more generally. Patients with cerebellar damage show impairment in card sequencing tests. In one such test, individual cards show cartoons, depicting characters that are actors in brief stories. The subjects have to arrange these cards in a meaningful way to create a logical story (Leggio et al., 2008). The literature describing the roles of cerebellar operations in language is rich. Let us focus here on nonmotor aspects of language alone. Cerebellar lesions are often associated with decreased verbal fluency—both, a difficulty finding words to continue or finish a sentence, and agrammatism (speech contains content words, but lacks function words) (Molinari et al., 1997; Schmahmann, 2004; Stoodley and Schmahmann, 2009; see also Hodge et al., 2009). Verbal fluency is of specific interest for these considerations. As argued by Marco Molinari and Maria Leggio in a contribution to a consensus paper on 'Language and the cerebellum'

(Marien *et al.*, 2014), 'fluency tasks are particularly valuable, because they assess associative processes— phonological and semantic—and strategic abilities in word searching'. In a semantic fluency task, for example, the subject is asked to name as many words as possible that belong to a semantic category, such as 'birds'. Impairment in such semantic fluency tasks is typical for cerebellar patients. Such studies show that the cerebellum plays a central role in the two aspects of language that I have pointed out before to highlight potential roles of associative plasticity: semantic fluency and word placement.

It is no accident that the cerebellum surfaced as one of the brain areas involved in these aspects of language. The cerebellum is one of the brain centers that support associative learning. In delay eyeblink conditioning, an experimental test for associative learning that I introduced in Chapter V (Figure 5.4), the CS is a predictor of the US signal. Therefore, learning the association between CS and US stimuli allows animals to generate a well-timed eyelid closure before an airpuff hits the otherwise unprotected eye. Typical CS–US intervals used in such experiments range from 150 to 1,000 ms. Not all cerebellar modules control motor output, though. Modules outside of the eyeblink control area may regulate communication with neocortical areas instead. Therefore, a more general view on what cerebellar associative learning does is required. All cerebellar circuits integrate complex incoming signals at the level of Purkinje cells and their dendrites, and create an output signal. In a general description of cerebellar operations, we can consider an output that is, after a short delay, followed by an error signal, as 'faulty'. The mere occurrence of the error signal will then lead to a depression of those parallel fiber (PF) synapses that contributed to the generation of this output. In contrast, if an output signal is interpreted as beneficial—by the absence of an error signal or otherwise—the PF synapses involved may undergo potentiation. We therefore can predict for a cerebellar involvement in nonmotor aspects of language that LTD may prevent the establishment of faulty associations such as the one between the words 'eat' and 'book.' Such an 'error-guided' *negative* association learning (detection and elimination of nonsensical associations) alone would constitute a critical contribution of the cerebellum to language acquisition.

However, the role of climbing fiber (CF) signaling in cerebellar associative learning may be even more complex. CF activity is likely not restricted to error signals alone. Instead, the occurrence of complex spikes in Purkinje cells—which is driven by CFs—is also observed in situations where harmless sensory stimuli are presented such as brief light pulses or auditory signals (Ohmae and Medina, 2015). As pointed out by these authors, the sensory CF signals are particularly reliable and strong when these signals are novel, but they do persist even after repetitive stimulus presentation. Like PFs, CFs may thus also present signals in an 'error-neutral' context. If this interpretation is correct, it has consequences for the polarity of synaptic gain change. Recent work from my own laboratory suggests that LTD only results when complex spikes with specific features are presented: these spikes need to occur in clusters of two or more complex spikes, together with the preceding PF signal covering a total activation period of about 200–400 ms (PF–CF–CF; Titley et al., 2019). PF activity paired with individual complex spikes leads to LTP instead. This finding is a departure from the classical scenario: that CF co-activity causes LTD, whereas the absence of CF signals causes LTP (Coesmans et al., 2004). In recordings from awake mice, we observed that the probability that complex spikes occur in clusters of two or more is inversely related to response latency (Titley et al., 2019). This finding suggests that response urgency—for example, when an error signal is detected—is associated with a higher probability for prolonged complex spike firing. The occurrence of individual versus clustered complex spikes may thus distinguish between complex spikes conveying harmless sensory stimuli and those conveying instructive error signals, respectively. This distinction enables the 'selection' of the right type of associative plasticity that is appropriate to either stabilize synapses (when simple associations of the style 'A' to 'B' are established) or weaken/eliminate synapses (when causal relationships with an error signal are detected).

The previous considerations have dealt with individual association 'steps', that is, single associations between 'A' and 'B.' A recent publication by Michael Mauk at the University of Texas at Austin shows that associative cerebellar learning may create sequences of conditioned behavior, too. In this case, the feedback signal from a completed behavior serves as a cue for learning a subsequent association (Khilkevich *et al.*, 2018). This important finding explains how the cerebellum may support the learning of entire sequences of associations.

## Language Impairment and Cerebellar Dysfunction in Autism

Deficits in the cellular machinery involved in language acquisition are clearly evident in autism. Some of these deficits are found in the cerebellum. In light of strong evidence of cerebellar dysfunction in autism, this should not come as a surprise. To begin with, there is a strong correlation between cerebellar damage (at birth) and the occurrence of autism (Wang *et al.*, 2014). Eyeblink conditioning is impaired in human autistic individuals (Sears *et al.*, 1994; Oristaglio *et al.*, 2013; Welsh and Oristaglio, 2016) as well as in mouse models of autism (Piochon *et al.*, 2014; Kloth *et al.*, 2015). In addition, cerebellar LTD—one of the cellular correlates of eyeblink conditioning—is impaired in autism mouse models (Table 5.1; Piochon *et al.*, 2016). Indeed, autism and cerebellar damage show overlap in specific symptoms related to language. Autistic children often have an impairment of nonmotor aspects of language, such as vocabulary and semantics (the meaning of words or sentences), as well as syntax (set of rules that govern the structure of sentences to create meaning) (Rapin, 1996; Tager-Flusberg, 2006). There is, however, a wider perspective on these findings that I would like to share. The computational power of our brains rests on the ability to detect and interpret meaning in the outer world, and to create meaning itself. Interpreting the world outside to 'read out' meaning includes—among more functions—the understanding of causal relationships and symbolic associations. This is the function of cellular mechanisms that support associative learning, maintained by the cerebellum and other brain areas. The observation that cerebellar associative learning is impaired in autism therefore can tell us a lot about the difficulties in interpreting the outer world—and making sense of it—that appear to be characteristic of autistic individuals.

# References

Abraham, W.C. (2003). How long will long-term potentiation last? *Phil. Trans. R. Soc. Lond. B* 358, 735–744.

Belmeguenai, A., Hosy, E., Bengtsson, F., Pedroarena, C.M., Piochon, C., Teuling, E., He, Q., Ohtsuki, G., De Jeu, M.T., Elgersma, Y., De Zeeuw, C.I., Jörntell, H., and Hansel, C. (2010). Intrinsic plasticity complements long-term potentiation in parallel fiber input gain control in cerebellar Purkinje cells. *J. Neurosci.* 30, 13630–13643.

Cai, D.J., *et al.* (2016). A shared neural ensemble links distinct contextual memories encoded close in time. *Nature* 534, 115–118.

Frick, A., Magee, J., and Johnston, D. (2004). LTP is accompanied by an enhanced local excitability of pyramidal neuron dendrites. *Nat. Neurosci.* 7, 126–135.

Gill, D.F., and Hansel, C. (2020). Muscarinic downregulation of SK2-type K+ conductances promotes intrinsic plasticity in L2/3 pyramidal neurons of the mouse primary somatosensory cortex. *eNeuro* 7, ENEURO. 0453-19.2020.

Hansel, C., Linden, D.J., and D'Angelo, E. (2001). Beyond parallel fiber LTD: The diversity of synaptic and non-synaptic plasticity in the cerebellum. *Nat. Neurosci.* 4, 467–475.

Hattori, S., Yoon, T., Disterhoft, J.F., and Weiss, C. (2014). Functional reorganization of a prefrontal cortical network mediating consolidation of trace eyeblink conditioning. *J. Neurosci.* 34, 1432–1445.

Hodge, S.M., Makris, N., Kennedy, D.N., Caviness, V.S., Howard, J., McGrath, L., Steele, S., Frazier, J.A., Tager-Flusberg, H., and Harris, G.J. (2009). Cerebellum, language, and cognition in Autism and specific language impairment. *J. Autism Dev. Disord.* 40, 300–316.

Khilkevich, A., Zambrano, J., Richards, M.M., and Mauk, M.D. (2018). Cerebellar implementation of movement sequences through feedback. *eLife* 7, e37443.

Kloth, A.D., *et al.* (2015). Cerebellar associative sensory learning defects in five mouse autism models. *eLife* 4, e06085.

Leggio, M.G., Tedesco, A.M., Chiricozzi, F.R., Clausi, S., Orsini, A., and Molinari, M. (2008). Cognitive sequencing impairment in patients with focal or atrophic cerebellar damage. *Brain* 131, 1332–1343.

Lopez-Rojas, J., Heine, M., and Kreutz, M.R. (2016). Plasticity of intrinsic excitability in mature granule cells of the dentate gyrus. *Sci. Rep.* 6, 21615.

Mahon, S., and Charpier, S. (2012). Bidirectional plasticity of intrinsic excitability controls sensory inputs efficiency in layer 5 barrel cortex neurons *in vivo*. *J. Neurosci.* 32, 11377–11389.

Marien, P., *et al.* (2014). Consensus paper: language and the cerebellum: an ongoing enigma. *Cerebellum* 13, 386–410.

McKay, B.M., Matthews, E.A., Oliveira, F.A., and Disterhoft, J.F. (2009). Intrinsic neuronal excitability is reversibly altered by a single experience in fear conditioning. *J. Neurophysiol.* 102, 2763–2770.

Molinari, M., Leggio, M.G., and Silveri, M.C. (1997). Verbal fluency and agrammatism. In R.J. Bradley, R.A. Harris, & P. Jenner (Series Eds.), J.D. Schmahmann (Vol. Ed.), *International Review of Neurobiology: Vol. 41. The Cerebellum and Cognition* (San Diego: Academic Press), pp. 325–339.

Moyer, J.R., Thompson, L.T., and Disterhoft, J.F. (1996). Trace eyeblink conditioning increases CA1 excitability in a transient and learning-specific manner. *J. Neurosci.* 16, 5536–5546.

Ohmae, S., and Medina, J.F. (2015). Climbing fibers encode a temporal-difference prediction error during cerebellar learning in mice. *Nat. Neurosci.* 18, 1798–1803.

Ohtsuki, G., Piochon, C., Adelman, J.P., and Hansel, C. (2012). SK2 channel modulation contributes to compartment-specific dendritic plasticity in cerebellar Purkinje cells. *Neuron* 75, 108–120.

Oristaglio, J., Hyman West, S., Ghaffari, M., Lech, M.S., Verma, B.R., Harvey, J.A., Welsh, J.P., and Malone, R.P. (2013). Children with autism spectrum disorders show abnormal conditioned response timing on delay, but not trace, eyeblink conditioning. *Neuroscience* 248, 708–718.

Pignatelli, M., Ryan, T.J., Roy, D.S., Lovett, C., Smith, L.M., Muralidhar, S., and Tonegawa, S. (2019). Engram cell excitability state determines the efficacy of memory retrieval. *Neuron* 101, 274–284.

Piochon, C., Kano, M., and Hansel, C. (2016). LTD-like molecular pathways in developmental synaptic pruning. *Nat. Neurosci.* 19, 1299–1310.

Piochon, C., Kloth, A.D., Grasselli, G., Titley, H.K., Nakayama, H., Hashimoto, K., Wan, V., Simmons, D.H., Eissa, T., Nakatani, J., Cherskov, A., Miyazaki, T., Watanabe, M., Takumi, T., Wang, S.S., and Hansel, C. (2014). Cerebellar plasticity and motor learning deficits in a copy-number variation mouse model of autism. *Nat. Commun.* 5, 5586.

Quiroga, R.Q. (2012). Concept cells: the building blocks of declarative memory functions. *Nat. Rev. Neurosci.* 13, 587–597.

Quiroga, R.Q., Reddy, L., Kreiman, G., Koch, C., and Fried, I. (2005). Invariant visual representation by single neurons in the human brain. *Nature* 435, 1102–1107.

Rapin, I. (Ed.). (1996). *Clinics in developmental medicine: Vol. 139. Preschool children with inadequate communication* (London: Mac Keith Press).

Rashid, A.J., Yan, C., Mercaldo, V., Hsiang, H.L., Park, S., Cole, C.J., De Cristofaro, A., Yu, J., Ramakrishnan, C., Lee, S.Y., Deisseroth, K., Frankland, P.W., and Josselyn, S.A. (2016). Competition between engrams influences fear memory formation and recall. *Science* 353, 383–387.

Salin, P.A., Malenka, R.C., and Nicoll, R.A. (1996). Cyclic AMP mediates a presynaptic form of LTP at cerebellar parallel fiber synapses. *Neuron* 16, 797–803.

Schmahmann, J.D. (2004). Disorders of the cerebellum: Ataxia, dysmetria of thought, and the cerebellar cognitive affective syndrome. *J. Neuropsychiatr. and Clin. Neurosci.* 16, 367–378.

Schonewille, M., Belmeguenai, A., Koekkoek, S.K., Houtman, S.H., Boele, H.J., van Beugen, B.J., Gao, Z., Badura, A., Ohtsuki, G., Amerika, W.E., Hosy, E., Hoebeek, F.E., Elgersma, Y., Hansel, C., and De Zeeuw, C.I. (2010). Purkinje cell-specific knockout of the protein phosphatase PP2B impairs potentiation and cerebellar motor learning. *Neuron* 67, 618–628.

Schreurs, B.G., Gusev, P.A., Tomsic, D., Alkon, D.L., and Shi, T. (1998). Intracellular correlates of acquisition and long-term memory of classical conditioning in Purkinje cell dendrites in slices of rabbit cerebellar lobule HVI. *J. Neurosci.* 18, 5498–5507.

Sears, L.L., Finn, P.R., and Steinmetz, J.E. (1994). Abnormal classical eyeblink conditioning in autism. *J. Autism Dev. Disord.* 24, 737–751.

Singer, W. (1995). Development and plasticity of cortical processing architectures. *Science* 270, 758–764.

Stoodley, C.J., and Schmahmann, J.D. (2009). The cerebellum and language: evidence from patients with cerebellar degeneration. *Brain & Lang.* 110, 149–153.

Tager-Flusberg, H. (2006). Defining language phenotypes in autism. *Clin. Neurosci. Res.* 6, 219–224.

Thompson, L.T., Moyer, J.R., and Disterhoft, J.F. (1996). Transient changes in excitability of rabbit CA3 neurons with a time course appropriate to support memory consolidation. *J. Neurophysiol.* 76, 1836–1849.

Titley, H.K., Brunel, N., and Hansel, C. (2017). Toward a neurocentric view of learning. *Neuron* 95, 19–32.

Titley, H.K., Kislin, M., Simmons, D.H., Wang, S.S., and Hansel, C. (2019). Complex spike clusters and false-positive rejection in a cerebellar supervised learning rule. *J. Physiol.* 597.16, 4387–4406.

Titley, H.K., Watkins, G.V., Lin, C., Weiss, C., McCarthy, M., Disterhoft, J.F., and Hansel, C. (2020). Intrinsic excitability increase in cerebellar Purkinje cells following delay eyeblink conditioning in mice. *J.Neurosci.* 40, 2038–2046.

Wang, S.S., Kloth, A., and Badura, A. (2014). The cerebellum, sensitive periods, and autism. *Neuron* 83, 518–532.

Welsh, J.P., and Oristaglio, J.T. (2016). Autism and classical eyeblink conditioning: performance changes of the conditioned response related to autism spectrum disorder diagnosis. *Front. Psychiatry* 7, 137.

Yiu, A.P., Mercaldo, V., Yan, C., Richards, B., Rashid, A.J., Hsiang, H.L., Pressey, J., Mahadevan, V., Tran, M.M., Kushner, S.A., Woodin, M.A., Frankland, P.W., and Josselyn, S.A. (2014). Neurons are recruited to a memory trace based on relative neuronal excitability immediately before training. *Neuron* 83, 722–735.

# 10 Plasticity and the Shaping of Individual Brains

In his masterful satirical farce *Mein Kampf*, playwright George Tabori places a young Adolf Hitler, an ambitious but untalented artist, into a flophouse in Vienna. There, he meets and befriends a Jew, Shlomo Herzl, who plans to write a book, *Mein Kampf*, while taking young Adolf under his wings. Shlomo—unaware what is to come—mocks the struggle of his protégé to evolve his future dictatorial worldview in light of their friendship:

> *"Forget Shlomo, ONE Jew, only talk about THE Jews, and you will be a king, who walks on a carpet of bones"*

In the animal kingdom, the appearance of differences among individuals from the same species likely presented an evolutionary advantage, allowing animals to probe morphological or behavioral variations against the fitness to produce viable offspring. In humans, individuality has a different meaning: it is an accepted social norm throughout most cultures that simply by being unique, by differing from any other human being, we gain the right to be looked at, to be assessed, and to be judged individually. Nowhere in the literature are the importance and the consequence of individuality better captured than in Tabori's dark comedy. So long as Hitler acknowledges Shlomo, or any other Jew as an individual, he cannot hatefully talk about them. To do that, he needs to de-personalize them, to replace the individual with the anonymous mass of people. ('THE Jews' was indeed one of Hitler's favorite words to bark into the microphones, in an attempt to spread blame and hate.) This importance of individuality in human society is the reason why in this concluding chapter I

167

will write about how the concept of engrams is important for the making of individual brains.

## Plasticity, Circuit Learning and the Origin of Individuality

The claim that every (sufficiently important) experience leaves a trace in our brain, and therefore shapes it, does not mean that all experiences equally foster individuality. Learning the sequence of signals that a traffic light uses to control safe passage is such a widely shared experience that it would hardly be justified to talk about it in the context of individuality. The same holds true for some, but not all, motor sequences, as well as for basic language skills (within the geographical region where your native tongue is spoken). Similarly, developmental synaptic pruning in many cases optimizes neural circuits for tasks that are so fundamental that there is little tolerance for individual variations. This is different—in the developing and the adult brain—for information content related to personal, nonessential experiences. Often, the processing and storage of those memories depends on prior experiences and emotions associated with these experiences.

I saw George Tabori's *Mein Kampf* performed at the *Schauspielhaus* Theater in Zurich when I was an undergraduate student at the University of Zurich. My parents were visiting, and we went to see it together. I remember that my father was quite uneasy with the play, which is not surprising, as he had experienced the terrors of the war as a little boy in Germany under the rule of the Nazis. For my father, it was harder, if not impossible, to look back at those times through the satirical lens that Tabori adopted. My father, my mother, and I shared the experience of seeing this play. It left traces in our brains that certainly are shared and somewhat similar, as they are similarly shared by the hundreds of others in the audience. However, this is not a memory shared by people who never saw the play, or did not see it performed in that production at the Schauspielhaus. Despite this shared experience, to each of us—my father, my mother, and myself—the play meant something different. Accordingly, the memories that each of us formed will likely be associated with different emotions and evaluations, and they will give rise to different subsequent thoughts and mind processes.

Assigning value to experiences and memories is an important aspect of learning. This step is possible, because of the tight anatomical and functional connections between the brain centers involved in cognitive and social tasks (hippocampus; neocortex) and the limbic system, the brain center of emotional control. The limbic system consists of several substructures, among them the hippocampus itself, the ventral tegmental area (VTA), and the amygdala. The VTA can be considered as—at a somewhat simplified level—a center for positive reactions, pleasure and reward, and it operates with dopaminergic signaling pathways. Conversely, the amygdala is a center that controls anxiety and fear (again, this is a rather simplified depiction). This is not the place to discuss how emotional control works *per se*, but it is important to point out that—in ways that are not understood in detail—the interactions between these pathways appear to result in the activity of connected ensembles that enable the recall of specific memories in some association with judgmental or emotional thoughts. Value assignment is fundamental to how we navigate our complex world: it informs decision-making, determines whether we engage in approach or avoidance behaviors, and guides our social interactions.

## Individuality and Self-awareness Arising from Experience

The uniqueness of our brains, and thus the basis of our individual personalities, results from our own experiences and the memory traces that they leave behind. This notion might be self-evident to some, and disconcerting to others. My father, who was deeply religious, would certainly have worried about this. Does this idea leave any space for divine creation? I believe that there is no definite answer that arises from the sciences. It appears that the ability of our brains to adapt, to store memories, and to enable individuality can be explained by known biological principles. What is not known seems in reach, at least when granting some level of extrapolation of scientific knowledge. At the same time, it is obvious to me that such biological plausibility does not at all exclude creative acts. This question remains one of faith. We are more likely to explain the nature of our uniqueness. I have argued in this book that our brains reflect organismal history, both a genetic history that is shaped by cross-generational experiences of selection pressures and a history of personal experiences. Our brains are shaped by past events. We know almost nothing about genetic influences on inner-species variations in the organization or function of neural circuits. That is why this highly interesting topic has only been marginally discussed in this book. We know more about engrams formed by personal experiences.

Before I turn to the nature of memory traces, let me briefly revisit the question of how we should talk about learning. I outlined in the Introduction that 'learning' is typically interpreted as *behavioral learning*. This is certainly how the word is used in ordinary speech and by many academic disciplines. I have contrasted this with *circuit learning*, a term that implies that any change to the *physical substance* of the brain in response to an experience or activity constitutes a form of learning. My arguments for this expansion of the definition (or use) of the term 'learning' are twofold. First, behavioral learning only allows for a rather course description of learning phenomena. This is particularly true for the study of learning in animal models, where a limited number of—typically rather simple and reductionist—behavioral tests are used in a laboratory environment. If we understand changes in brain matter as learning, we achieve a higher 'graininess' of description, and with that, more accuracy. Second, in my view, the behaving organism in its entirety is still an accumulation of matter, very much like—though bigger and more complex—a specific brain circuit. Both receive input and generate output that can be altered during learning. This view in part constitutes a departure from behaviorism as developed by the American psychologist B.F. Skinner. Behaviorism looks for measurable outcomes that it seeks to find in behavioral output. At first approximation, this school of thought is incompatible with the focus on circuit learning championed here. At the same time, and quite ironically so, an emphasis on circuit learning can also be seen as a *radical form of behaviorism*. That is because in this view components of learning are dissected into parts that are supported by individual brain circuits. Behavior itself is not required any longer at this level of description, but now every brain circuit, in fact every synapse and neuron, becomes its own 'learning center', reinforcing structural and functional changes that result in 'good output' while suppressing those that cause 'bad output.'

Circuit learning via experience leaves a physical trace. If the experience is personal, that is, not shared by others, this memory trace is unique and contributes to individuality (i.e. the history of experiences that defines us as individuals is reflected in such personal memory traces). The memory trace can be a morphological change at a microscale level, such as a change in the density or shape of dendritic spines (Engert and Bonhoeffer, 1999; Trachtenberg *et al.*, 2002; Hofer *et al.*, 2009). Morphological changes of this nature certainly take place during developmental synaptic pruning. However, LTP, LTD and other forms of cellular plasticity are not always associated with structural spine alterations (Sdrulla and Linden, 2007; Thomazeau *et al.*, 2020). Physical changes may also occur at an even smaller, nanoscale level: LTP and LTD rely on changes in the density of glutamate receptors. This addition or removal of receptors is ultimately a physical change, as much as $K^+$ channel trafficking in intrinsic plasticity is a physical change. Changes in the density and shape of spines, as well as changes in the membrane expression of transmitter receptors and ion channels, constitute one major component of a physical memory trace. A second component is the content of stored information. A synaptic change, as in LTP, ensures that specific information content is processed. The identity of this content thus becomes 'engraphed'—as Richard Semon put it—into brain matter. Specific information content engraphed via microscopic structural changes in the brain: this is possibly the best way to understand memory, and with it the origins of uniqueness.

In conclusion, I would like to note that my attempt to deduct individuality from both organismal history and from personal experiences can also be extended to the perception of self. The Austrian philosopher Karl Popper has pointed out that self-awareness is the result of both species-typical genetic dispositions and personal, particularly social, experiences. According to Popper, we are not born with a sense of self, but it is acquired through learning.[1] Before acquiring language, a child must have the experience of being called by its name and receiving attention. The experience of this directed attention enables a distinction of oneself from others, and thus the realization that there is a 'self.' I generally agree with Popper on this notion. However, Popper did not write about how experience also contributes to the very existence of individuality in the first place. This is the aspect of *self-formation* that this book describes. Memory traces create uniqueness in neural circuits and entire brains, and unlike individuality resulting from our inherited genome,[2] this uniqueness is absolute.

# References

Engert, F., and Bonhoeffer, T. (1999). Dendritic spine changes associated with hippocampal long-term synaptic plasticity. *Nature* 399, 66–70.

Hofer, S.B., Mrsic-Flogel, T.D., Bonhoeffer, T., and Hubener, M. (2009). Experience leaves a lasting structural trace in cortical circuits. *Nature* 457, 313–317.

Sdrulla, A.D., and Linden, D.J. (2007). Double dissociation between long-term depression and dendritic spine morphology in cerebellar Purkinje cells. *Nat. Neurosci.* 10, 546–548.

Thomazeau, A., Bosch, M., Essayan-Perez, S., Barnes, S.A., De Jesus-Cortes, H., and Bear, M.F. (2020). Dissociation of functional and structural plasticity of dendritic spines during NMDAR and mGluR-dependent long-term synaptic depression in wild-type and fragile X model mice. *Mol. Psychiatry*, doi.org/10.1038/s41380-020-0821-6.

Trachtenberg, J.T., Chen, B.E., Knott, G.W., Feng, G., Sanes, J.R., Welker, E., and Svoboda, K. (2002). Long-term in vivo imaging of experience-dependent synaptic plasticity in adult cortex. *Nature* 420, 788–794.

---

[1] Popper, K., and Eccles, J. (1977). *The Self and Its Brain* (Berlin, London, New York: Springer Verlag Heidelberg).

[2] Identical twins are genetically the same, thus they are not truly unique based on their genome alone.

# Index

www.ingramcontent.com/pod-product-compliance
Lightning Source LLC
Chambersburg PA
CBHW050628190326

41458CB00008B/2178